UNITS, DIMENSIONAL ANALYSIS AND PHYSICAL SIMILARITY

Units, Dimensional Analysis and Physical Similarity

B. S. MASSEY
Reader in Mechanical Engineering
University College London

VAN NOSTRAND REINHOLD COMPANY
LONDON
NEW YORK   CINCINNATI   TORONTO · MELBOURNE

VAN NOSTRAND REINHOLD COMPANY
Windsor House, 46 Victoria Street, London, S.W.1

INTERNATIONAL OFFICES
New York   Cincinnati   Toronto   Melbourne

Copyright © 1971 B. S. MASSEY

All rights reserved. No part of this publication may be reproduced, stored in a retrieval system, or transmitted, in any form, or by any means, electronic, mechanical, photocopying, recording, or otherwise without the prior permission of the copyright owner.

Library of Congress Catalog Card No. 74–160201
ISBN: 0 442 05178 6

First Published 1971

Printed in Great Britain by
Butler & Tanner Ltd, Frome and London

Preface

This book deals with matters which are basic to the whole of physical science and engineering. Yet they are matters which tend to be overlooked or, at any rate, inadequately treated. The teaching of all branches of science and engineering is divided and subdivided into innumerable fragments, and, necessary though concentration on individual fragments may be, it too often prevents appreciation of the nature of the whole. Both the clock-conscious teacher and the examination-conscious taught are concerned to draw from the library of wider knowledge only such information as will enable them to take the next step along their narrow specialist path. So connections between different specialisms are often not seen and the applicability of universal laws is not appreciated.

Nowadays it is seldom that units, dimensional analysis and so on are considered outside the context of particular subjects of study. Consequently many questions are not treated at all and general principles are hardly hinted at. Indeed, students seem to be expected to perceive many of these matters by light of nature. But the light of nature is sometimes dim, and one of my chief aims in writing this book has been to illuminate those nooks and crannies which nature leaves dark.

Much of the book's purpose will be achieved if it leads the reader to acquire an instinctive feeling for 'physical algebra' and for arguments based on dimensional formulae. It is not a philosophical treatise. It does not recount history, although occasionally historical reasons are brought in to explain otherwise inexplicable oddities. For that matter, in discussing units, for instance, I have not always followed the historical development, nor have I used the concepts on which original definitions were based if later thinking has shown those concepts to have been ill-founded.

I have made no attempt to recount the disagreements, historical or contemporary, between those of differing views. In any case, many arguments of this kind prove to be more semantic than philosophic. Words are often used with different meanings by different people, or even by the same person at different times. I have therefore taken some trouble to define terms carefully, and I hope that readers will pay due heed to the definitions. The subject of dimensional analysis in particular has frequently been presented as a kind of mystic hocus-pocus, a set of occult recipes to be blindly followed. Much of its power and value have thus not been realized. Here, with no more than the minimum of mathematics, I have tried to bring precision, system and clarity where these qualities have too often been lacking.

Although I have aimed at usefulness rather than encyclopaedic completeness, there is one instance—the list of dimensionless parameters in Section 7.4—where completeness and usefulness can hardly be separated. The only deliberate omissions from the list are those parameters which have been suggested but not, it seems, brought into use. Even so, the achievement doubtless falls short of the intention, partly because such information is acquired largely by serendipity but also because new parameters now seem to be christened almost daily. I should therefore welcome additional information from readers.

Material in this book comes from so many sources that I cannot hope to acknowledge them all individually. I am much indebted to many writers whose names, alas, I mostly forget but whose ideas have fertilized my own thinking. I am grateful too to those students who, over several years, have let me submit many notions and techniques to the test of their understanding. And it is a pleasure to acknowledge here the valuable help I have had from many colleagues who have allowed me to sharpen my wits on theirs, and who, with remarkable patience, have responded to my requests for their expert guidance. Of these I should like to mention especially Dr J. A. Barnard, Dr J. W. Fox, Mr C. E. Gimson, Prof. K. J. Ives, Mr H. Marriott and Prof. J. W. Mullin. To them this book owes much. Yet I hasten to absolve all these friends from any share in the imperfections of my work: these I claim as my own. It is also a pleasure to record my sincere appreciation to Miss C. R. Clewer for her sure-fingered typing of the manuscript, to Dr B. R. Clayton for his assistance as a volunteer proofreader, and to the staff of the publishers for their unfailing help and encouragement.

B. S. M.

Contents

PREFACE

CHAPTER 1 INTRODUCTION   1
    1.1 The Philosophy of Measurement   1
    1.2 'Physical Algebra'   2
    1.3 Scalar and Vector Quantities   4
    1.4 Symbols for Vector Quantities   5

CHAPTER 2 NEWTON'S DEFINITIONS AND LAWS OF MOTION   6
    2.1 Newton Interpreted   6
    2.2 Newton's Second Law in Modern Mechanics   9
    2.3 Mass and Weight   10

CHAPTER 3 UNITS—THEIR NATURE AND MAGNITUDE   13
    3.1 Base Units and Derived Units   13
    3.2 The Principal Units of Mechanics   14
        3.2.1 Units of Length   14
        3.2.2 Units of Mass   15
        3.2.3 Unit of Time Interval   16
        3.2.4 Units of Force and Other Units of Mass   16
        3.2.5 Differences between U.S. and 'Imperial' Units   20
        3.2.6 Summary   21
    3.3 Some other Units of Mechanics   22
    3.4 Units for Thermal Quantities   22
        3.4.1 The Thermodynamic Definition of the Magnitude of Temperature   24
        3.4.2 Practical Scales of Temperature   28
        3.4.3 Units for Quantity of Heat   29
    3.5 Units for Electric and Magnetic Quantities   30
        3.5.1 The c.g.s. Electrostatic System of Units   31
        3.5.2 The c.g.s. Electromagnetic System of Units   31
        3.5.3 The Practical System of Units   33
        3.5.4 The International MKSA System of Units   34
        3.5.5 Rationalized Equations   35
    3.6 Units for Quantities connected with Illumination   36

CHAPTER 4 THE SYSTÈME INTERNATIONAL D'UNITÉS   38
    4.1 Introduction   38
    4.2 Prefixes and Conventions in Printing   39
    4.3 Units for Pressure and Stress   41

CHAPTER 5 CONVERSION FACTORS   42

CHAPTER 6 THE FORM OF EXPRESSIONS IN PHYSICAL ALGEBRA   48
    6.1 Types of Magnitudes and Equations   48
    6.2 The Nature of Dimensional Formulae   50

		6.3	Dimensional Homogeneity	52
		6.4	Dimensional Analysis	54
			6.4.1 The Process of Analysis	55
			6.4.2 Rayleigh's Method	56
			6.4.3 The 'Pi' Theorem	63
		6.5	Dimensional Formula of Plane Angle	72
		6.6	Dimensional Formulae for Thermal Quantities	75
		6.7	Dimensional Formulae for Electric and Magnetic Quantities	79
			6.7.1 Dimensional Formulae of Magnitudes of Electric Quantities	79
			6.7.2 Dimensional Formulae of Magnitudes of Magnetic Quantities	81
			6.7.3 Examples of Dimensional Analysis with Electric and Magnetic Quantities	82
		6.8	Closing Remarks on Dimensional Analysis	86
Chapter 7	Physical Similarity			89
		7.1	Introduction	89
		7.2	Types of Physical Similarity	90
			7.2.1 Geometric Similarity	90
			7.2.2 Kinematic Similarity	90
			7.2.3 Dynamic Similarity	91
			7.2.4 Other Kinds of Similarity	91
		7.3	Mathematical Expression of Similarity Requirements	92
		7.4	Named Dimensionless Parameters	93
Appendix 1	Symbols, Dimensional Formulae and Units for Principal Physical Quantities			123
Appendix 2	Approximate Values of some Physical Constants and Common Properties			128
Appendix 3	Some Other Units used in Scientific Work			130
Further Reading				133
List of Unit Symbols				134
Index				137

1

Introduction

1.1 The Philosophy of Measurement

All measurement is essentially comparison. Consider for example the measurement of length. Any given length can be specified only by stating the number of times that it contains some other given length—known as the unit of length—which is taken as standard. Thus, to measure the length of a table one compares the length of the table with some 'unit length' such as one inch. Suppose that the unit length has to be used 36 times in succession to make a length equal to that of the table. Then the length of the table is 36 times the length of the unit, i.e. $36 \times (1 \text{ inch})$, or more briefly, 36 inches.

A statement of the length of the table therefore necessarily consists of two parts: (1) the 'numeric', that is, the number of times that the unit must be successively used to equal the length of the table, and (2) the unit itself.

We have just taken the measurement of length as an example. Length is one kind of physical quantity—that is, something which can be measured by some strictly definable process. In principle, the measurement of any physical quantity consists of comparison with a standard amount of that quantity, and the standard amount is termed the unit. The result of the measurement is known as the magnitude of the quantity.

We shall see later that, for many quantities, direct comparison with the unit is made only with great difficulty, if at all. Yet whether the comparison is direct or indirect, any statement of the magnitude must consist of the two parts, the numeric and the unit. The magnitude Q of any measurable quantity is expressible in the general form

$$Q = nU \tag{1.1}$$

where n represents the numeric and U the unit.

Let us consider again the measurement of the length of a table. The inch is only one of a number of units which might be used for the purpose. The metre might be used, or the foot, or the cubit, or the width of a chair. But whether the length of the table were quoted as 36 inches, 0·9144 metres, 3 feet, 1·71 cubits or 2·12 chair-widths, the actual length of the table would remain the same. No magnitude is itself in any way altered by the method used to determine it. True, the numeric may be different: the result of a measurement is expressed, however, not by the numeric alone but by the product of the numeric and the unit. Or, to use again the algebraic form of eqn 1.1:

$$Q = n_1 U_1 = n_2 U_2 = n_3 U_3 = \text{etc.}$$

Employing a different U simply entails a correspondingly different n. The magnitude Q is unaltered.

1.2 'Physical Algebra'

The algebra used in describing physical causes and effects is not the same as the algebra of pure mathematics, even though it has the same superficial appearance. The algebra of pure mathematics (which we may refer to as 'ordinary algebra') is essentially the expression of relations among *numbers* (even though various symbols are usually employed to represent the numbers). These numbers are not numbers of anything in particular; in fact, to what *things* the numbers may relate is not the concern of the pure mathematician at all.

'Physical' algebra, on the other hand, is a means of expressing relations among the magnitudes of physical quantities—force, velocity, mass, energy, and so on. It tells us how the magnitude of one quantity depends on the magnitudes of others. In 'physical' algebra†, therefore, most of the symbols represent not numbers but the magnitudes of physical quantities.

Now this fact places certain restrictions on the algebra. The mathematical processes of addition, subtraction and equating have intelligible meaning only when applied to magnitudes of quantities of the same kind. We cannot add or equate a mass and an interval of time, or a mass and a velocity, but only one mass and another mass, one time interval and another time interval, and so on.‡ When therefore we have a relation between the magnitudes of physical quantities —in the form of a mathematical equation for example—the question arises whether both sides of the equation represent magnitudes of the same kind of quantity. If physical quantities are of the *same kind*, then their magnitudes may be expressed in terms of the *same unit*. Therefore one condition which a relation must fulfil if it is to be correct and to have real meaning is that any terms which are added, subtracted or equated must represent magnitudes which may be expressed in terms of the same unit. When this condition is met, the equation is said to be *dimensionally homogeneous*. (This requirement will be discussed further in Section 6.3.)

Any relations in 'physical' algebra, then, must possess a twofold consistency. There must first be the normal numerical relation of 'ordinary' algebra: in an equation, for example, the two sides must be equal in magnitude. But besides this, and indeed of primary importance, a relation in 'physical' algebra must have a uniform *character*: any terms which are added, subtracted or equated must refer to quantities of the same kind. Whereas a relation in 'ordinary' algebra is basically a means of comparing numbers—no matter how the numbers are obtained or to what things they refer—a relation in 'physical' algebra is a means of comparing the magnitudes of similar quantities. Thus terms of a

† Some writers have used the term 'quantity calculus' for this concept.

‡ One may *write*, for example, 5 metres + 3 minutes—just as one may write 5 apples + 3 chairs—but the process of addition cannot be performed and it is not possible to proceed further. Here, then, the plus sign is an empty promise and thus meaningless. Even the statement '5 apples + 3 chairs = 8 *things*' represents no real progress because the right-hand side lacks precision, and the statement, although valid when read from left to right, is not necessarily true when read from right to left, as a genuine equation would be. Indeed, the 'equals' sign would be better written $\Rightarrow$ to indicate the direction in which the statement is valid.

§1.2]   'PHYSICAL ALGEBRA'

'physical' relation which are added, subtracted or equated must represent the magnitudes of quantities of the same physical nature.

At this point we may usefully mention one or two consequences of the fact that terms which are added must be of the same kind. One consequence is that the arguments of mathematical functions which are expressible as power series must be simply numerics with no units. For example

$$e^x = 1 + x + \frac{x^2}{2!} + \frac{x^3}{3!} + \frac{x^4}{4!} + \cdots$$

Unless x is only a numeric, the terms in the series cannot all be of the same kind, that is the series cannot be dimensionally homogeneous. Thus, if x does have units, to write e^x is meaningless.

If x is simply a numeric, then all the terms in the series are numerics and $y = e^x$ must also be a numeric. From this it follows that the equivalent statement $x = \ln y$ is similarly restricted. Logarithms may be taken only of numbers. (True, there are careless writers who present expressions including terms such as $\ln r$, where r represents a radius, that is a length. It will usually be found, however, that the $\ln r$ has arisen from the integration of dr/r and that the writer omitted the integration constant. In such a case the $\ln r$ term should be $\ln r - \ln r_0 = \ln (r/r_0)$, where $-\ln r_0$ is the missing constant.)

Similar considerations apply to the various trigonometrical and hyperbolic functions. Sin x, cos x, tan x, sinh x, tanh x and so on can each be expressed as a series of powers of x multiplied by numerical coefficients—but only if the x is a numeric.

It is also worth remarking on the physical nature of differential coefficients and integrals. By definition,

$$\frac{dy}{dx} = \lim_{\Delta x \to 0} \left(\frac{\Delta y}{\Delta x}\right)$$

where Δy represents the increment of y corresponding to Δx, the increment of x. Thus dy/dx, $\Delta y/\Delta x$ and y/x are all of the same physical nature. The corresponding second-order differential coefficient d^2y/dx^2 is defined as

$$\frac{d}{dx}\left(\frac{dy}{dx}\right)$$

and so has the same physical nature as

$$\frac{\text{increment of } dy/dx}{\text{increment of } x} \quad \text{or as} \quad \frac{dy/dx}{x}$$

that is the same nature as

$$\frac{y/x}{x} = \frac{y}{x^2}$$

For example, if s represents a distance and t an interval of time, then ds/dt has the nature of a speed and d^2s/dt^2 has the nature of an acceleration.

The integral $\int y \, dx$ is the limit, as $\Delta x \to 0$, of the sum of the products $y \, \Delta x$, each of which is similar in nature to yx. For example, if F represents the magnitude of a force, and x the magnitude of a distance in the same direction as the force, then $\int F \, dx$ represents an amount of (force × parallel distance), that is, an amount of work or energy.

Many of the difficulties and uncertainties encountered in calculations concerning physical phenomena stem from a failure to recognize the vital distinction between 'physical' algebra and 'ordinary' algebra. The symbols in a 'physical' formula refer not just to numerical magnitudes but to the magnitudes of physical quantities; not just to numerics but to the numerics and the units together; the formula is not symbolic arithmetic but symbolic physics or engineering. This being so, the truth of a formula in 'physical' algebra does not depend simply on the numbers which may be associated with the symbols.

1.3 Scalar and Vector Quantities

Many physical quantities, such as time interval, density or temperature, are completely specified by their magnitudes. Such quantities are termed *scalar quantities*, or, more simply, *scalars*. There are other quantities, however, which cannot be completely specified unless the information given includes not only the magnitude of the quantity but also a statement of direction or orientation.

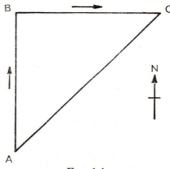

Fig. 1.1

Examples of such quantities are force, velocity and linear displacement; these are known as *vector quantities*, or, more simply, *vectors*.

Strictly, a vector is a straight line used as a geometrical representation of a vector quantity. By the adoption of some suitable scale factor, the length of the line is made proportional to the magnitude of the quantity. An arrow head on the line may then be used to indicate the *sense* of the direction (that is, to indicate, for example, whether a north–south line is to be regarded as pointing to the north or to the south). However, this verbal distinction between a vector and a vector quantity is now not often observed.

Confusion between scalar and vector quantities sometimes arises when different names are used to refer to closely related concepts, but the names are not strictly synonymous. It is then important to distinguish between the different names. One instance in which the distinction is important is the pair of names *distance* and *linear displacement*. If a man at a point A (see Fig. 1.1) walks 1 kilometre due north over level ground to the point B and then walks 1 kilometre due east (again over level ground) to the point C, he traverses a total *distance* of 2 kilometres. His final position, however, is the same as if he had walked directly from A to C in one straight line. The position of C is $\sqrt{2}$ kilometres north-east of A, and so the *linear displacement* of this hypothetical pedestrian is $\sqrt{2}$ kilometres N.E. Even if he took such a devious route that he traversed 100 kilometres before arriving at C his linear displacement would still be $\sqrt{2}$ kilometres N.E. from A.

The direction 'N.E.' must be specified in order to distinguish the point C from any other on the circumference of a circle of centre A and radius $\sqrt{2}$ kilometres. A complete description of linear displacement thus requires a statement not only of the distance of the terminal point from the starting point, but also the direction in which that distance must be measured. Linear displacement is therefore a vector quantity whereas distance is generally regarded as a scalar quantity.

Two other terms which are not quite synonymous are *velocity* and *speed*. In conventional use, 'velocity' describes a vector concept, that is, it is concerned not only with rapidity of movement but also with the direction in which the movement takes place. On the other hand, 'speed' conventionally refers only to the magnitude of the velocity and leaves the direction unspecified.

In modern use, the term 'vector quantity' is further restricted to quantities which also obey certain rules in 'vector algebra'. 'Directional' quantities which do not obey all these rules are sometimes known as *pseudovectors*. Such a quantity cannot be represented uniquely by a straight line with an arrow head on it unless some additional convention is adopted. This is true, for example, of quantities involving rotation: the straight line may represent the axis about which rotation takes place, but whether the rotation is clockwise or anticlockwise cannot be specified by an arrow *along* the line unless some further convention is used. Quantities which necessarily imply rotation about an axis (for example, angular velocity, moment of force) are pseudovectors. So is area: the direction of an area may be represented by a line perpendicular to it, but a convention is needed to specify which side of a plane, for example, is considered to face in the positive sense of the direction. The present custom, therefore, is to reserve the term 'vector quantity' for those quantities which may be represented completely by a line of definite length and an arrow along it.

Confusion, or at least uncertainty, sometimes arises over the word 'direction'. In Fig. 1.1, for example, the line AC represents the *line of action* of the linear displacement. The addition of an arrow head to this line then indicates whether the linear displacement is from A to C or from C to A. The arrow points in the *sense* of the direction: a movement from C to A is said to be of opposite sense to a movement from A to C. Some writers use the word 'direction' to refer only to the line of action; the sense then needs separate statement. More usually, however, the sense is included in the meaning of the word 'direction'.

1.4 Symbols for Vector Quantities

In mathematical work, it is often desirable to distinguish vector quantities from scalar quantities by the use of symbols of a different kind. Printers commonly do this by using bold-face (heavy) type for vector quantities: for example, **F** to represent force, but m to represent mass, a scalar quantity. This method of distinguishing vector from scalar quantities is hardly feasible in handwriting, and so an arrow is then often placed immediately above the symbol for a vector quantity, thus: $\vec{F}$.

2

Newton's Definitions and Laws of Motion

2.1 Newton Interpreted

All modern systems of measurement—that is, the means by which the magnitudes of physical quantities are expressed—make use of the definitions and laws of motion published in 1687 by Sir Isaac Newton (1642–1727). Brief consideration of some of these definitions and laws is therefore desirable before we examine systems of measurement, especially as some features of Newton's work are often imperfectly understood.

His work was published in Latin, but his Third and Fourth Definitions may be rendered thus:

All matter has the property of *inertia*, whereby every body (that is, every piece of matter), free from outside influences, persists in its state either of rest or of uniform motion in a straight line.

A *force* is an action exerted on a body, tending to change its state either of rest or of uniform motion in a straight line.

These two definitions lead to Newton's First Law:

Every body persists in its state of rest or of uniform motion in a straight line, except in so far as it is compelled to change that state by forces acting on it.

Or, in rather more detail:

If at any instant the velocity of a body is changing—in magnitude or direction or both—then at that instant the body is acted upon by a net force from outside the body. The body itself tends to remain in a state either of rest or of uniform motion in a straight line. This tendency is called the inertia of the body, and inertia is a characteristic of all matter.

This had previously been recognized by Galileo (1564–1642) in 1638, and so Newton's First Law has sometimes been known as Galileo's Law of Inertia.

A force of a particular kind is that which, in the absence of any restraint, causes a body to move towards the earth. This force of attraction exerted by the earth on a body was realized by Newton to be an example of the force of attraction to be found between any two pieces of matter anywhere. All matter, irrespective of its chemical composition or its physical state, thus has two properties. One is *inertia*, that is the tendency for the velocity of the body (which may of course be zero) to remain unchanged. The other is the capacity of any piece of matter to attract towards it any other piece of matter. This is the property of *gravitational attraction*.

Both the inertia of a body and the gravitational attraction which it exerts depend on the quantity of matter which constitutes that body. A measure of the quantity of matter is therefore of great importance in discussing the effects of inertia and gravitational attraction. Thus the concept of *mass* is required. The mass of a body may be defined simply as a measure of the quantity of matter which comprises the body. This definition, although for the moment lacking the complete precision which we shall bring to it in Section 2.3, allows the further definition of *momentum* (Newton's Second Definition):

The momentum of a body is that quantity whose magnitude is given by the product of the magnitudes of the velocity of the body and its mass.

From this we may proceed to Newton's Second Law:

The rate of increase of momentum of a body (that is, the increase of momentum divided by the infinitesimal time interval during which it takes place) is proportional to the net force acting on the body, and takes place along the direction in which that force is applied.

The specifying of direction here emphasizes that both force and momentum are vector quantities.

It might be thought that Newton's First Law does not need separate statement because it forms a special case of his Second Law. It is, however, more than this. In any measurement the zero of the scale must be defined. The statement 'Edinburgh is 390 miles' is meaningless: the words 'from London' must be added to define the starting point of measurement, that is, the zero of the scale. Similarly in the measurement of time: a man told to rise at 7 hours must know whether the number of hours is to be measured from the time he goes to bed or —as is more usual—from midnight. And just as we cannot define 'motion' until we have defined a state of 'rest' or 'no motion' with which motion may be compared, so it is necessary to define the state of 'no force' before the effect of force itself may be discussed. Newton's First Law, far from being merely a special case of his Second (as has sometimes been thought), has the important function of defining the state of 'no force'. In fact, the First Law might be rendered:

If at any instant the velocity of a body is *not* changing (either in magnitude or in direction), then at that instant no net force from outside the body is acting on it.

Thus the First Law defines the zero of the scale of force. Only when this has been done can the Second Law have any real meaning.

This definition of the state of 'no force', however, depends on the definition of 'no change in velocity'. The zero of the scale of velocity therefore also needs consideration.

Measurements of length (that is, the shortest distance between two points) may be made with reference to a set of co-ordinate axes (for example, Ox, Oy and Oz). These axes provide the zeros of the scales of measurement for the x, y and z co-ordinates respectively, and thus constitute a 'frame of reference' for the measurements. Velocity, that is rate of increase of linear displacement, is also measured with respect to such a frame of reference. In addition, of course, some kind of clock must be used to measure time interval. Similarly, measurements of acceleration must be made with respect to the frame of reference and the clock.

When Newton formulated his laws the frame of reference he had in mind was one which was stationary with respect to the so-called fixed stars. The frames which we choose for measurement purposes, however, are most often stationary with respect not to the fixed stars but to the earth. The zero of the force scale which we now accept is determined by reference to a frame stationary with respect to the earth. Sometimes, however, a frame of reference is moving relative to the earth: we may wish to make measurements on a moving ship or railway train, for example. The question therefore arises: How, if at all, does the movement of the frame affect the measurements?

Newton's Laws truly describe physical phenomena only if the necessary measurements are made with respect to a suitable frame of reference. A frame for which Newton's Laws are valid is known as an *inertial frame of reference*. In an inertial frame, bodies with no net force acting on them move in straight lines at constant velocity or stay at rest; net forces give rise to rates of increase of momentum proportional to the forces. A frame stationary with respect to the fixed stars is an inertial frame; one moving with uniform constant velocity is also an inertial frame. However, an accelerating frame is a *non*-inertial frame because a body stationary *with respect to that frame* must itself have an acceleration and consequently a net force acts on the body; for an accelerating frame, therefore, Newton's First Law is invalid. A rotating frame is an example of an accelerating frame because all rotary motion involves acceleration towards the axis of rotation.

It can be argued that the spinning earth is thus not a perfect inertial frame. The magnitude of the centripetal acceleration at points on the surface of the earth is, however, small enough to be negligible for most practical purposes. Phenomena observed in earthly laboratories agree very closely with Newton's Laws, and a reference frame fixed with respect to the earth is usually regarded as one to which final appeal may be made. But such close agreement with Newton's Laws could not be expected in a laboratory set up in a tossing ship, for example, or in a motor car rounding a corner at speed.

Newton's Second Law, then, is valid only for an inertial frame of reference—that is, one not itself accelerating. (In what follows, inertial frames will be assumed unless the contrary is stated.)

We now briefly consider Newton's Third Law. It is frequently stated in the following form: To every action (i.e. force) there is an equal and opposite reaction. However, to make it clear that the forces of action and reaction *do not act on the same body*, a better statement of the law is:

If a body A exerts a force on a body B, then body B exerts an equal and opposite force on body A.

A force is not something existing, as it were, in its own right. Every force is exerted by one body on another body. Which of the two bodies is regarded as supplying the action and which the reaction is immaterial. Furthermore, action and reaction remain equal and opposite even when the bodies concerned are undergoing acceleration.

A falling stone, for example, attracts the earth just as strongly as the earth attracts the stone. A train always pulls on its locomotive just as strongly as the locomotive pulls on the train—whether or not the velocity is constant. Or think of a tug-of-war team winning a pull. The winning team exerts a total force of magnitude F_1 on the rope, and the rope exerts a total force of magnitude F_1 on them. At the other end of the rope similar conditions apply: the other team exerts a total force of magnitude F_2 on the rope, and the rope exerts a total force of magnitude F_2 on them. If F_1 does not equal F_2 the rope has a *net* force of magnitude $F_1 - F_2$ acting on it. Thus the rope (and probably both the teams) will move. But even then, the force with which either team pulls on their own end of the rope must exactly equal the force with which the rope pulls on them.

2.2 Newton's Second Law in Modern Mechanics

The first part of Newton's Second Law may be reduced to the following 'shorthand' form:

$$\text{Force} \propto \text{Rate of increase of momentum}$$

The second part—the statement that the increase of momentum is in the same direction as the force—must not be forgotten, but it is the first part to which we now give particular attention.

Consider a body, the mass of which (that is, the 'quantity of matter') is constant, and which moves so that, at any instant of time, all parts of the body have the same velocity. For this case we may write

$$\text{Force} \propto \frac{\text{Increase of (mass} \times \text{velocity)}}{\text{Corresponding infinitesimal time interval}}$$

$$= \text{Mass} \times \frac{\text{Increase of velocity}}{\text{Corresponding infinitesimal time interval}}$$

$$= \text{Mass} \times \text{Acceleration}$$

The acceleration, which is the vector part of the product, takes place in the same direction as the force.

So far we have been able to say what *kind* of quantity force is, but we have had no means of specifying the magnitude of a force. In fact, it is impossible to specify the magnitude of a force accurately and directly without reference to other quantities. In other words, to specify the magnitude of a force we need some formula expressing the relation between that magnitude and the magnitudes of other quantities. Yet all such relations—like Newton's Second Law—are necessarily in the form of proportionalities until the magnitude of a force can be specified.

It is therefore necessary to select one of these proportionalities and convert it into an equation. The particular relation chosen to define the magnitude of a force is Newton's Second Law. The coefficient of proportionality has been arbitrarily fixed at unity, and our entire modern system of measurement has been based on the relation

$$\text{Force EQUALS Mass} \times \text{Acceleration}$$

As we shall see later, this modern form of Newton's Second Law, $F = ma$, holds no matter how we measure the quantities force F, mass m and acceleration a.

The form $F = ma$ is the one in which Newton's Second Law is most often used. It should be remembered, however, that it applies only to those instances in which the mass of the body in question is constant and the acceleration of all parts of the body is the same.

2.3 Mass and Weight

Mass has been defined as quantity of matter, or, as Lewis Carroll's Dormouse would perhaps have said, 'a muchness'. It is something having magnitude but not direction: that is, a scalar quantity. Now although the definition 'quantity of matter' makes possible a kind of intuitive understanding of the idea and may even serve as a philosophic definition, it is insufficient as a practical one for it lacks quantitative precision. It still leaves us, after all, with the task of defining 'quantity of matter'.

To arrive at a more precise definition we make use of Newton's Third Law. Let us consider two isolated bodies, that is two bodies so far away from any third body that the latter has no influence on them. Neither of the bodies has any electric or magnetic effect. If each of these two isolated bodies (say '1' and '2') exerts a force on the other, these forces must be equal in magnitude (although opposite in direction). Hence, by the Second Law, since the individual 'quantities of matter' are constant, the products of mass and acceleration for each body are identical in magnitude. That is, if m represents mass and a acceleration,

$$m_1 a_1 = -m_2 a_2$$
$$\therefore m_1/m_2 = -a_2/a_1$$

The negative sign arises because the accelerations, like the forces, are opposite in direction. Therefore either a_1 or a_2 is negative. If only the magnitudes ($|a_1|$ and $|a_2|$) of the accelerations are considered, then

$$m_1/m_2 = |a_2|/|a_1|$$

In other words, the ratio of the masses is the inverse ratio of the magnitudes of the accelerations produced by equal forces.

This principle is utilized in the ballistic balance for comparing masses. Such a balance, however, is not normally used for this purpose because a far greater accuracy may be obtained from a beam balance.

It was recognized by Galileo—if not earlier—that all bodies in the same locality undergo the same downward acceleration in free fall (strictly, that is, in a vacuum, but in practice with negligible air resistance). The *weight* of a body which is prevented from falling may be defined as the force which the body exerts vertically downwards on its supports. A support may be something on which the body rests (for example, the platform of a weighing machine) or it may be the suspension from which the body hangs (for example, the string of a plumb-bob). We may ourselves act as a support: the heaviness of an object held in the hand is experienced as the force which it exerts on the hand. This force known as weight arises from the gravitational attraction between the body and the earth, although the gravitational attraction has another effect which we shall examine in a moment. The support provides a reaction to the weight, and so the body is in equilibrium under equal and opposite forces.

If, however, the support is withdrawn, the body is no longer in equilibrium, and the unbalanced force then causes the body to accelerate towards the earth. If the body is of mass m and is in a locality where the acceleration of bodies in free fall is g, then the force acting on the body is given by the Second Law as the product of the mass and the acceleration:

$$W = mg \qquad (2.1)$$

Here W represents the magnitude of the force formerly acting on the support, that is the weight of the body.

The equation shows that it is possible to compare the masses of two bodies by comparing their weights, provided that the comparison is made under conditions in which the acceleration would be unchanged. The beam balance provides a ready means of such comparison by arranging that the forces necessary to support the bodies are equal. As the value of g is the same for the two bodies in question its actual magnitude is unimportant.

The comparison is usually made between a mass of unknown magnitude and the mass of a reference body—or a selection of reference bodies. These reference

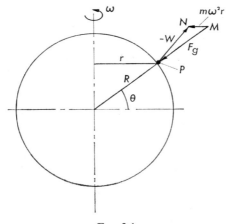

Fig. 2.1

bodies are (unfortunately) usually termed 'weights'. Here is but one instance of how confusion between mass and weight arises in everyday language. Yet a clear distinction between the two quantities is vital for scientific purposes.

On the surface of the earth there is no significant difference between weight and the force of gravitational attraction. We may, however, spend a moment considering the distinction between these two forces in view of the interest nowadays in man-made satellites and the condition of 'weightlessness' experienced by the travellers in them. Figure 2.1 shows a point P at latitude θ on the surface of the earth (of radius R). As the earth rotates on its axis with angular velocity ω the point P moves in a circle of radius $r = R \cos \theta$. Now a body at P experiences a gravitational attraction F_g towards the centre of the earth. If the body is stationary relative to the earth the necessary reaction force provided by its support is equal and opposite to the weight W. The reaction force may therefore be represented by $-W$. There is, however, a small angle (never greater than 0·1°) between the lines of action of F_g and of $-W$. These two forces therefore have a resultant, and it is this resultant which provides the necessary acceleration of magnitude $\omega^2 r$ towards the axis of rotation. If the body is of mass m the corresponding force is $m\omega^2 r$.

At the surface of the earth the force $m\omega^2 r$ is very small, and so F_g and W are substantially the same. Indeed, at the north and south poles $m\omega^2 r$ is zero and F_g and W are there identical. A satellite, however, has a larger angular velocity (ω_s) and the radius of its orbit (r_s) is somewhat larger than the radius of the earth. Thus for a body in a satellite $m\omega_s^2 r_s$ is not small and in some circumstances $m\omega_s^2 r_s$ and F_g are identical. Then W (the force exerted by the body on its supports) is zero and the body is said to be weightless. In fact, the weight W may be regarded as the force 'left over' after $m\omega^2 r$ has been taken from the force of gravitational attraction. In a satellite in orbit all of F_g is used in providing $m\omega_s^2 r_s$, and so no force remains as weight.

When a body falls freely the only force acting on it is F_g and it therefore falls in the direction of F_g (i.e. MP in the diagram). To an earth-bound observer, however, who himself has an acceleration $\omega^2 r$ towards the axis of the earth's rotation, only the relative acceleration is apparent: it seems to him that the fall follows the direction NP and that a force W is causing it. Thus, so long as we are not concerned with phenomena associated with satellites or similar circumstances we may regard each body close to the earth as acted on by a force of appropriate magnitude W—the weight of the body—tending to pull it towards the earth.

It has become customary to say that the symbol g represents the acceleration due to gravity. This custom is in many respects unfortunate. The acceleration referred to is that of a body falling freely (strictly therefore in a vacuum), conditions which form a very limited field of study. In any case this acceleration is not something fundamental: it is simply an effect produced by the weight of the body. Any body, whether falling or not, experiences the earth's gravitational attraction. Of this attraction, that component known as the weight of the body will, if no resistance is offered to it, give the body an acceleration g, the magnitude of which may be expressed by eqn 2.1 as $g = W/m$.

Now the magnitude of the force F_g is determined by Newton's Law of Universal Gravitation and is directly proportional to the mass of the body. The force $m\omega^2 r$ is also directly proportional to the mass of the body. Thus the force W, which completes the triangle of forces on the body, is directly proportional to the mass. Although W is also dependent on the position of the body relative to the earth, in the same locality the ratio W/m (that is, g) is constant. It is, then, much more logical to regard the symbol g as representing the *weight per unit mass*.

3

Units—Their Nature and Magnitude

3.1 Base Units and Derived Units

We have seen that to express the magnitude of a physical quantity we require not only a number ('numeric') but also an appropriate unit. A unit of a quantity is a suitable sample of that kind of quantity. An inch or a millimetre, for instance, is a sample of the kind of quantity known as length. The size of these samples or units is arbitrary. For example, an arbitrary length is used as the unit of length, an arbitrary mass as the unit of mass, an arbitrary interval of time as the unit of time. If a unit is to be defined in terms of some material object serving as the arbitrary standard, then this standard will necessarily be unique, and actual practical measurements must be made by using copies of the standard. It must therefore be possible for any material standard not only to be preserved but also to be faithfully reproduced.

An arbitrary length and an arbitrary mass do fulfil these conditions. Bodies having a definite length and bodies of a definite mass may be accurately made from suitable materials; they can be preserved without change; and exact copies may be made for general use.

In the measurement of time interval any phenomenon which regularly repeats itself may be used: the process of measurement consists of counting the repetitions. For centuries the motions of the planets have served as the phenomena to which reference is made. An arbitrary time interval may be specified from astronomical observations, and it may be accurately reproduced by pendulum experiments. Recently the vibration of atoms of caesium in the so-called 'atomic clock' has been used.

Units, such as those just mentioned, are termed *base* (or *fundamental* or *primary*) units because they are defined simply with reference to some standard, and not in terms of any other units. However, there are many quantities—velocity and acceleration, for example—for which material standards cannot be preserved, and, in any case, comparison with the standard would be very difficult, if not impossible. Units of such quantities are therefore defined in terms of the base units and are known as *derived* (or *secondary*) units.

Base units, such as those of length, mass and time interval, have no special significance apart from serving as the starting point from which the rest of the system of units is derived. Base units are chosen for reasons of practical convenience. However, every system of units for mechanics is in fact based on the magnitudes of a standard length, a standard mass and a standard interval of time.

In other fields of study further base units may be called for. These will be considered in Sections 3.4–3.6.

3.2 The Principal Units of Mechanics

A unit, as we have seen, is a standard magnitude of a particular kind of quantity; a sample of the quantity to which, quite arbitrarily, is assigned the numeric 'one'. The sizes of base units are determined not by laws of nature but by laws of governments. For although the choice of a unit is arbitrary it is clearly in the interests of honest commerce and ease of communication that the same units should be employed by most people for each kind of quantity. Therefore governments decide on the base units to be used in their countries, and arrange for the construction and preservation of material standards with which all other measures may be compared.

In the course of history many base units have been adopted in various countries, and from these base units vast numbers of other units have been derived. In this book we shall consider only those which have had wide currency in modern times. Later, in Chapter 4, we shall deal with the system of units which, it is hoped, will soon supersede all others.

For precise measurements, the only units which are now strictly fundamental—in the sense that they are not defined in terms of any other units—are those of what is loosely termed the 'metric system'. This system arose from the desire of the French Government, just after the French Revolution of 1789, for a single set of units in place of the great and confusing variety of local units which had by then come into being in different parts of France. Even traditional British (and American) units such as the foot and the pound mass have been redefined in terms of the corresponding metric units.

3.2.1 Units of Length

The first unit of the metric system to be defined was one of length, and this unit is termed the *metre* (abbreviated 'm'). The original definition of the metre was 10^{-7} times the distance from the north pole to the equator along the meridian which passes through Paris. Standard metre lengths were then marked on accurately constructed metal bars. This definition of the metre had the virtue of using a permanent measure provided by nature rather than some man-made reference standard which might be lost or damaged. However, the original determination of the distance from pole to equator was later found to be somewhat inaccurate, and the International Committee of Weights and Measures agreed not to adjust the size of the metre to give an exact ratio of 10^{-7}, but to redefine the metre as the distance between two marks on a standard metal bar under exactly specified conditions.

This bar, known as the International Prototype Metre, was used as the standard of length until 1960. It was kept by the International Bureau of Weights and Measures at Sèvres, near Paris, and a number of exact copies were made, one of which is still in the National Physical Laboratory at Teddington, Middlesex.

The desirability of employing a natural rather than a man-made standard prompted the suggestion, as long ago as 1828, that the wavelength of light should be used. Since 1960, therefore, the metre has been redefined in terms of the wavelength corresponding to a particular line in the spectrum of light produced by an electric discharge through the krypton isotope Kr-86.

The metre is now being increasingly used as the base unit of length in most countries for purposes of trade as well as science.

Ideally, units should be of such a size that the magnitudes expressed in terms of those units do not involve numerics which are inconveniently large or small. However, since no single unit for a particular quantity is likely to be of suitable size for all measurements, provision has to be made for units both smaller and larger than the basic unit for that quantity. In the 'metric system', the size of any of these smaller or larger units is related to that of the corresponding basic unit by a power of ten, and, in general, the name given to one of these units consists of a prefix plus the name of the basic unit.

The original prefixes covered the range 10^{-3} to 10^3; others have since been added and the list is now as follows.

Prefix	Symbol	Numerical factor by which basic unit is multiplied
tera	T	10^{12}
giga (pronounced *jy'-ga*)	G	10^9
mega (pronounced *meg'-ga*)	M	10^6
kilo	k	10^3
hecto	h	10^2
deca	da	10
deci	d	10^{-1}
centi	c	10^{-2}
milli	m	10^{-3}
micro	μ	10^{-6}
nano	n	10^{-9}
pico (pronounced *pee'-ko*)	p	10^{-12}
femto	f	10^{-15}
atto	a	10^{-18}

Thus a centimetre, for example, is 10^{-2} metre and has the abbreviation 'cm'. One thousand metres is one kilometre (km).

In Great Britain, the Commonwealth and the U.S.A. another standard unit of length is often used. This is the 'International Standard Yard', defined as 0·9144 metre precisely. Units for other quantities are more usually derived from the foot (abbreviated 'ft') which is one-third of the International Standard Yard. (Until 1959 the standard unit for all purposes in Great Britain and the Commonwealth was the 'Imperial Standard Yard'. This was defined as the distance between marks on a bar of bronze kept at the Board of Trade in London. The difference between the Imperial Standard Yard and the International Standard Yard is, for practical purposes, negligible.)

3.2.2 UNITS OF MASS

When the metric system was devised, the unit of mass was chosen as the mass of a cube of pure water, at normal atmospheric pressure, and at the temperature at which water has its maximum density. Apparently a cube of one metre side was thought to provide a mass unit too large for most purposes, and a cube of one centimetre side was decided on instead. The resulting mass unit was named the *gramme*, now often spelt *gram* in English-speaking countries.

In practice, a block of metal is more convenient as a reference standard, but a block of metal with a mass of one gram is rather small. So a one kilogram block, which is 1000 times larger, was used and this was made to correspond with the mass of one cubic decimetre of pure water at a standard temperature and pressure. It was later discovered that the cube used to contain the water was slightly too large. Rather than adjust the size of the kilogram, it was decided in 1875 to make a new block of metal, equal in mass to that already in use, and to use the new block as the fundamental standard. This new standard, termed the International Prototype Kilogram, is a cylinder of platinum–iridium alloy, and a number of exact copies have been made, one of which is kept at the National Physical Laboratory.

In Great Britain, much of the Commonwealth and the U.S.A. the International Standard Pound Mass has been taken as the standard unit, and this is defined as 0·453 592 37 kilogram precisely. (Before 1959, the British unit was the avoirdupois pound mass, the mass of a cylinder of platinum called the Imperial Standard Pound kept at the Board of Trade. The difference between this unit and the International Standard Pound Mass is entirely negligible for practical purposes.)

3.2.3 Unit of Time Interval

As a basic unit of time interval the mean solar second is universally adopted. Astronomical observations define the mean solar day, which, by successive subdivision, may be used to determine the standard second. Although the astronomical observations can be made with a high degree of accuracy, even better results are obtainable when the definition of the standard second is based on the resonance vibration of atoms of the caesium isotope, Cs-133. Since 1967, therefore, the standard second has been defined in terms of this atomic standard. The internationally agreed abbreviation for second is 's', but the older abbreviation 'sec' is still widely used.

3.2.4 Units of Force and Other Units of Mass

From the three base units—those of length, mass and time interval—it is possible to define units of all other quantities in mechanics. We shall pay particular attention, however, to the definition of units of force and of other units of mass, for it is here that much confusion has arisen in the past.

If the metre is adopted as the unit of length and the second as the unit of time interval, the resulting unit of velocity is the metre per second (m/s), and the unit of acceleration is the (metre per second) per second. The latter may be abbreviated 'm/s^2', or 'm s^{-2}'. (The first form is perhaps preferable, if only for the reason that small minus signs are easily overlooked. However, the use of negative exponents leads to greater neatness in more complicated expressions.) Since force is quantitatively defined as mass × acceleration, a unit of force may be directly derived from the kilogram mass (abbreviated 'kg') as the unit of mass, the metre as the unit of length, and the second as the unit of time interval. If a body of mass 1 kg is to be given an acceleration of 1 m/s², the net force required to do this is determined by the equation $F = ma$ as

$$1 \text{ kg} \times 1 \text{ m/s}^2 = 1 \text{ kg m/s}^2$$

This unit of force kg m/s² is given the name 'newton' (abbreviated 'N'). (Notice that, although the unit is named after a person—Sir Isaac Newton—, when the word is used for a unit a small, not a capital, 'n' is used.)

In the 'c.g.s.' system of units, formerly almost universal among physicists, the unit of length is the centimetre (cm), the unit of mass the gram mass (g—but the abbreviation 'gm' was formerly much more common), and the unit of time interval the second (s). The derived unit of force is therefore g cm/s², which is termed the 'dyne'.

British and American practice has favoured the foot (ft) as the unit of length, the pound mass (in this book abbreviated as 'lbm') as the unit of mass, and the second as the unit of time interval. The unit of force directly derivable from these units is the lbm ft/s², to which the name 'poundal' (abbreviated 'pdl') has been given.

Although the poundal might seem to have much to commend it as a unit of force, it has in fact been little used, at any rate explicitly. One reason for its neglect is undoubtedly that instruments for measuring forces (such as spring balances) have never been calibrated in terms of poundals. The unit predominantly used has been the *pound-force*, and to understand its popularity (and also that of the corresponding unit, kilogram-force, used in much of Europe) we must digress a little.

The force by far the most commonly measured is weight. Yet, although what is measured is the force called weight, the purpose of the measurement is almost always to indicate the *mass*. In the commerce of the market-place and shop-counter goods are 'weighed', but what the purchaser is concerned with is the quantity of matter which he is buying: if he buys butter, for example, he is interested in the amount of nourishment to be obtained from it. And the nourishment in a pound of butter depends on its mass, not on the force tending to pull it towards the earth. It is natural, however, to speak of the butter as 'weighing one pound'.

Although a housewife is concerned with the amount of butter, or its 'muchness' (to quote the Dormouse again), in deciding how much she should buy as a week's supply for the family and whether she is getting value for money, she will be aware of the weight of the butter when carrying her purchases home. In such everyday affairs, mass and weight are inevitably closely linked—so closely, in fact, that it is rare for any distinction to be made between the two ideas.

It is therefore natural that a unit *of the same name* is used in everyday life for both mass and force. To give two things the same name is not of course to make them identical: it is merely confusing. The way in which spring balances and similar devices are calibrated is designed to minister to this duality.

Thus a unit which may be called the 'pound weight' has come into currency. It may be defined by saying that a stationary body of 1 lbm exerts on its supports a vertical force of 1 pound weight. An alternative statement of the magnitude of this force may be derived from the definition of g as weight/mass: 1 pound weight = 1 lbm × g.

As a unit of force, however, this quantity 1 pound weight is unsatisfactory because it is variable in magnitude. The vertical force necessary to support objects of a given mass varies slightly over the earth's surface (there is a difference of about 0·5 per cent between poles and equator) and varies more significantly with distance from the surface. The variation of this force is more conveniently expressed as a variation of g, the weight per unit mass.

For most purposes in the past such variations were of small importance and the pound weight was sufficiently constant to serve as a unit of force. However, for precise work the dependence of the magnitude of any unit on the location at which it is to be used cannot be tolerated. Consequently a unit termed the *pound-force* (abbreviated 'lbf') was defined which does not vary with location. A standard value of g was chosen—32·1740 ft/s^2—and 1 lbf equals the weight which a body of 1 lbm would exert at a place where g has this standard value.†

The magnitude of 1 pound-force is therefore 1 lbm × 32·1740 ft/s^2. Since this may also be written as 32·1740 lbm ft/s^2 it is identical with 32·1740 poundals. The numeric 32·1740 has more accuracy than is required for normal purposes and the figure 32·2 is generally used in calculations.

The use of a 'pound' as a unit of mass and another sort of 'pound' as a unit of force is a source of considerable confusion. It cannot be too strongly urged that the word 'pound' should never be used without the qualifying word 'mass' or 'force'. Similarly the abbreviations lbm and lbf are strongly recommended to supplant entirely the indeterminate 'lb'.

There is no fundamental reason why the magnitudes of physical quantities should be defined in a particular order. For example, if the magnitude of force can be satisfactorily specified to begin with, we can just as well use Newton's Second Law $F = ma$ to define the magnitude of mass m as F/a as to define the magnitude of force F as ma.

So by specifying the pound-force as a unit of force a derived unit of mass may be defined as 1 lbf ÷ 1 ft/s^2. This unit has been termed the *slug*. The name was chosen as a reminder that it is associated with the sluggishness or inertia of matter that is measured by mass. The advantage of using this unit lies in the fact that the numerics involved in its definition are each 1·0 and we therefore have a set of units in which a net force of 1 unit causes a body of mass 1 unit to move with an acceleration of 1 unit. (A set of units which are related to each other by equations in which no factor other than 1·0 appears is often known as a *coherent* set.) One slug may be defined as the mass of that body which is given an acceleration of 1 ft/s^2 by a net force of 1 lbf.

The slug is, however, merely a multiple of the pound mass, to which it is inevitably linked through the definition of the pound-force:

$$1 \text{ slug} = \frac{1 \text{ lbf}}{1 \text{ ft/s}^2} = \frac{1 \text{ lbm} \times 32 \cdot 1740 \text{ ft/s}^2}{1 \text{ ft/s}^2} = 32 \cdot 1740 \text{ lbm}$$

The definition of the slug in terms of the pound-force is thus seen to be really

† This standard value of g was first used about 1900 and was then chosen as the value at sea-level at 45° latitude. More accurate measurements since then have shown the figure to be slightly in error as an expression of the weight per unit mass at this location. To which precise point on the earth's surface the standard value corresponds is, however, immaterial and the value now universally accepted as the standard is 9·806 65 m/s^2 precisely, i.e. 32·1740 ft/s^2 approximately.

§3.2] THE PRINCIPAL UNITS OF MECHANICS 19

a circumlocution. (It is, in fact, doubtful whether a satisfactory definition of a unit of force could ever be achieved independently of a unit of mass.) The disadvantage of the slug as a unit of mass is that the mass of a body or its density (i.e. the ratio of mass to volume) has seldom been quoted in terms of the slug, and it can hardly be said that in Great Britain the slug has found wide acceptance.

If neither the slug nor the poundal is employed, use of the set of units lbm, ft, s, lbf requires particular alertness in the calculator. One may not decide to use lbf as the unit of force and lbm as the unit of mass, and then, in applying Newton's Second Law, expect the product of mass and acceleration—say (x lbm) times (y ft/s^2)—to be xy lbf. It will be xy lbm ft/s^2, i.e. xy pdl. Since 1 lbf $=$ 32·2 pdl (approximately), the result expressed in lbf is xy lbf/32·2 $=$ (xy/32·2) lbf.

There is a parallel history of confusion in the older versions of the 'metric system'. Units which should properly be called 'gram-force' (gf) and 'kilogram-force' (kgf) have been widely used in place of the dyne. These units need to be distinguished from the gram weight and kilogram weight, the magnitudes of which depend on geographical location. The general adoption, in continental countries, of units of the *Système International* (S.I.) (see Section 4.1), in which the kilogram is always the unit of mass and the newton the unit of force, is thus much to be welcomed.

A very dangerous practice which should be guarded against is the use of the letter g as an abbreviation for the numerics 9·806 65 or 32·1740. The letter g is the symbol for weight/mass; on the surface of the earth this quantity has the value (approximately) 9·806 65 N/kg $=$ 9·806 65 m/s^2. Consequently g may stand for 9·806 65 m/s^2; it may not also stand for the number 9·806 65. Or it may stand for 32·1740 ft/s^2 but not for 32·1740.

It is not too much to say that errors of units in calculations arise from the fundamental mistake of regarding 'physical algebra' as 'numerical algebra'. Physical algebra concerns magnitudes as a whole, not just numerics, and so the *units and numerics together* should be substituted for the symbols in formulae. The experienced calculator may put the numerics on paper and keep the units in his head, but it is hardly worth saving ink at the expense of anxiety and error.

Two other units of mass need mention here. For specifying the mass of a very small particle such as an atom or a molecule, a very small unit is used known as the *atomic mass unit* (a.m.u.), sometimes termed the *dalton*. Originally the mass of an ordinary hydrogen atom was taken as the unit; then one-sixteenth of the mass of an oxygen atom was used. However, when isotopes of oxygen were discovered, chemists preferred an atomic mass unit which was one-sixteenth of the average mass of all oxygen atoms, whereas physicists favoured a unit defined in terms of the oxygen-16 isotope only. Since 1960, however, these differences have been settled by agreement on a unit corresponding to one-twelfth of the mass of an atom of carbon-12. Accordingly, the unit fixed by this last definition is often termed the unified atomic mass unit. (The reason for the successive changes of definition has been to make the mass of as many isotopes as possible equal to an integral number of atomic mass units. The effect on the size of the unit has been slight.) The unified a.m.u. $\simeq$ 1·6604 $\times$ 10^{-27} kg.

It is interesting to note that the a.m.u. is defined in terms of a natural, not a man-made, standard. However, this natural standard, the mass of a carbon-12 atom, is not at present suitable for defining the kilogram because the ratio between the a.m.u. and the kilogram has not been determined with sufficient accuracy.

To refer to the amount of a substance taking part in a chemical reaction, chemists often find it convenient to use a unit proportional to the mass of a molecule of that substance. This is because, in any chemical change, substances react with one another in definite numbers of molecules so as to form definite numbers of other molecules. For instance, in the reaction

$$H_2SO_4 + 2NaOH = Na_2SO_4 + 2H_2O$$

one molecule of sulphuric acid reacts with two molecules of sodium hydroxide to form one molecule of sodium sulphate and two molecules of water. In practice, of course, the chemist is concerned with far more than a few molecules, and so expresses the amounts of the reactants in terms of a unit called the *mole* (abbreviated 'mol'). Although a more general definition will be given later, we shall for the moment define the mole as the product of 1 gram mass and the 'relative molecular mass' (often known as the 'molecular weight') of the substance, that is the ratio of the mass of the molecule to one-twelfth of the mass of an atom of carbon-12. One mole of oxygen (O_2), for example, is approximately 32 grams; one mole of hydrochloric acid (HCl) is approximately 36·5 grams.

Hence, for all pure substances, one mole comprises the same number of molecules. This number is expressed by the Avogadro constant, $6·023 \times 10^{23}$ mol^{-1}.

Other units used for the same purpose have been the *kilogram-mole* or *kilomole* (abbreviated 'kmol'), equal to the product of 1 kilogram mass and the relative molecular mass; and the *pound-mole*, equal to the product of 1 pound mass and the relative molecular mass.

Since the mole always contains the same number of molecules, the physical quantity of which the mole is the unit is simply the number of molecules. This physical quantity is now termed 'amount of substance' and its magnitude could in principle (though hardly in practice) be determined by counting the molecules. The *concept* of amount of substance thus does not depend on that of any other physical quantity. In other words, it is a fundamental concept.

However, in the present state of scientific knowledge, the accurate definition of the unit, the mole, has to be in terms of the relative molecular mass (which is simply a ratio) and a unit of mass such as the gram. The mole is thus in practice a derived unit. In fact it may be regarded as a unit of mass, the magnitude of which varies from substance to substance.

The concept of amount of substance has recently been widened to refer to particles other than molecules. The corresponding definition of the mole which will probably receive international agreement in 1971 is:

The mole is the amount of substance which contains as many elementary entities as there are atoms in 12 grams of carbon-12. (The elementary entities must be specified and may be atoms, molecules, ions, electrons, other particles, or specified groups of such particles.)

3.2.5 DIFFERENCES BETWEEN U.S. AND 'IMPERIAL' UNITS

Finally, there are some differences between U.S. and 'Imperial' units to which it may be well to draw attention. The most important concern measures of volume. The U.S. gallon is defined as 231 inch³ precisely, whereas the Imperial (British) gallon is defined as the volume occupied by 10 lbm of water under specified conditions of temperature and pressure and so is equivalent to 277·420 inch³. Thus the U.S. gallon is only about 83 per cent of the Imperial gallon. Other units of volume, such as bushels and 'dry gallons', also differ between the two countries, but these units are not likely to arise in scientific and engineering applications.

A multiple of the pound is the ton—both as a unit of mass and as a unit of force. In Great Britain the ton is 2240 pounds: in the U.S.A., Canada and South Africa the 'short ton' of 2000 pounds is usual, although the 'long ton' of 2240 pounds is used to some extent, especially in the American coal industry. Neither of these should be confused with the metric 'tonne' (abbreviated 't') which is 1000 kilograms, that is about 2205 pounds. The U.S. 'hundredweight' has 100 pounds in contrast to the British hundredweight of 112 pounds. (Both are abbreviated 'cwt'.)

3.2.6 SUMMARY

At this point we may summarize the definitions of the principal units of mass and of force.

1 kilogram (kg)	The *mass* of the International Prototype Kilogram—a certain block of metal.
1 newton (N)	That *force* which would give an acceleration of 1 m/s² to an unrestrained body of mass 1 kilogram.
1 gram (g)	One thousandth of the *mass* of the International Prototype Kilogram.
1 dyne (dyn)	That *force* which would give an acceleration of 1 cm/s² to an unrestrained body of mass 1 gram.
1 gram-force (gf)	= 980·665 dynes.
1 pound mass (lbm)	= 0·453 592 37 times the *mass* of the International Prototype Kilogram.
1 poundal (pdl)	That *force* which would give an acceleration of 1 ft/s² to an unrestrained body of 1 lbm.
1 pound-force (lbf)	= 32·1740 pdl.
[1 pound weight (lbwt)	is the downward vertical force which a stationary body of 1 lbm exerts on its supports. *This force varies with the location of the body.*]
1 slug (a unit of *mass*)	= 32·1740 lbm.

An unrestrained body of mass 1 slug would be given an acceleration of 1 ft/s² by 1 lbf.

Finally, three units of volume are defined.
1 litre = 10⁻⁵ cubic metres.

(This is the present exact definition of the litre. It was formerly defined as the volume occupied by 1 kilogram mass of pure water at 4°C and standard atmospheric pressure. The difference between the two definitions is about three parts in 10⁵. For measurements of high precision, the use of the word 'litre' is therefore discouraged. The agreed abbreviation for 'litre' is 'l', but, as this is easily mistaken for '1' (one), it is usually safer to write 'litre' in full.)

1 Imperial gallon is the volume occupied by 10 lbm of pure water under specified conditions (which are too detailed to concern us here).
1 U.S. gallon = 231 inch³.

3.3 Some Other Units of Mechanics

There is little point in referring here to units for every quantity which occurs in mechanics. We shall, however, consider a few units for which some words of explanation or clarification seem desirable.

A force moving along its own line of action does *work*. This is the quantity whose magnitude is given by the product of the force and the distance through which the point of application of the force moves. (If a force moves in some direction other than that of its own line of action, then only the component of the force in the direction of the movement does work. The amount of work done is the product of the distance moved and the component of the force in that direction—and this component may, of course, be negative.) The usual units are the metre newton, termed joule† (J), and the ft lbf.

Energy is defined as the capacity for doing work, and consequently its magnitude has the same units as work.

Power is the rate at which work is done. Its units are therefore those of work/time; for example, the joule/second, termed the watt (W), and the ft lbf/s. One horse power (abbreviated 'hp') is defined as 550 ft lbf/s. (The metric horse power, or 'cheval-vapeur', is defined as 75 metre kilogram-force per second and is therefore about 0·986 'British' horse power.)

Torque, or *moment of a force*, is a measure of the ability of a force to produce rotation. The magnitude of a torque is given by the product of the force and the perpendicular distance from its line of action to the instantaneous axis of rotation. The usual units are the newton metre (N m) and the lbf ft.

It will be noticed that, whereas the 'British' unit of work or energy is written ft lbf, that of torque is written with the components in the reverse order (lbf ft). This is done to distinguish the units of the two quantities work and torque which are fundamentally different.‡ Work is determined by the product of a force (or component of a force) and the distance *along its own line of action*: torque, on the other hand, is the product of a force (or a component) and a *perpendicular* distance.

3.4 Units for Thermal Quantities

For expressing the magnitudes of quantities related to heat, units derived only from those of length, mass and time interval are insufficient. An additional base unit is needed, and for centuries the quantity chosen to provide this extra unit has been temperature.

The fundamental notion of temperature comes from the sensation of warmth or coldness which we experience when we touch an object. This sensation is the

---

† Pronounced *jool*. Again notice that, although the unit is named after a person—the English physicist James Prescott Joule (1818–89)—, the word does not have a capital letter.
‡ The need for a similar distinction does not arise when the newton is used as the unit of force and the metre as that of length. This is because the metre newton is distinctively named 'joule'. In the c.g.s. system of units, the centimetre dyne (a unit of work or energy) is distinctively named 'erg'.

result of heat being transferred between the observer and the object he touches. The temperature of a body is in fact a measure of its ability to communicate heat to other bodies. Common experience further tells us that when two bodies (which are not equally hot or cold) are allowed to communicate heat to each other, the hotter body cools and the colder body becomes warmer until the two become equally hot or cold. We then say that the two bodies have reached the same temperature.

To indicate the temperature of a body a thermometer is used. This is essentially a small body T which can be placed so as to receive heat from the body A whose temperature is to be determined. After a sufficient time interval, T and A come to the same temperature and a measurement is then made of some suitable property of T which depends on the temperature. This property may be, for example, the volume of a definite amount of mercury held in a definite solid container, the length of a definite piece of metal, the electrical resistance of a definite piece of wire, the pressure exerted by a definite amount of a definite gas confined to a definite volume, and so on. The material in the thermometer must of course withstand the temperatures at which it is to be used, and return to its original thermal state after use. Also, the measured property must be one which varies continuously with temperature (that is, it must not cease to vary at any temperature in the range considered).

A temperature *scale* may be constructed by assigning, to particular values of the measured property, numbers which correspond to particular temperatures. Let us take as an example the familiar mercury-in-glass thermometer. This was largely the invention of the German physicist Gabriel D. Fahrenheit (1686–1736), and it was he who was responsible for the definition of a temperature scale based on two fixed points, that is on two standard temperatures. These two standard temperatures have usually been the melting point of ice and the boiling point of water, each at standard atmospheric pressure. The temperature scale may then be defined by making equal increments on it correspond to equal changes in the property being measured (e.g. equal movements of the mercury level in the glass stem of the thermometer).

The two standard temperatures are given standard numbers (e.g. 0 and 100); unfortunately, however, the number corresponding to any other temperature depends on the substance used in the thermometer and on the way in which its measured property changes with temperature. For instance, if the numbers were chosen with reference to the length of a tungsten rod, and the change in length between the two standard temperatures were divided into 100 equal parts, each corresponding to one 'degree' of temperature, then 50° would be the temperature at which the length was half-way between that for 0° and that for 100°. According to this 'tungsten-expansion scale', however, the 'half-way' length for a copper rod would be found at 54°. Moreover, a scale (that is, a set of numbers indicating temperature) so chosen that 'degrees' of temperature corresponded to equal changes in the electrical resistance of a copper wire would agree quite closely with the 'tungsten-expansion scale', and yet a 'tungsten-resistance scale' would be quite different. The question 'which property of which substance provides an equally divided scale?' has little meaning: we can arbitrarily choose any property to define the temperature scale and then claim that that scale is equally divided.

This difficulty illustrates the fact that temperature is, in an important respect, a quantity different in kind from length, mass, or time interval. A particular length, for example, may be measured by comparing it with a unit length applied in succession as many times as necessary. But one cannot apply unit temperature interval a number of times in succession to measure any other temperature.

3.4.1 The Thermodynamic Definition of the Magnitude of Temperature

Ideally, the definition of the numbers expressing temperature should not depend on a property of a particular substance. William Thomson, later Lord Kelvin (1824–1907), saw that Carnot's theory of a perfect reversible engine could be used to define the magnitude of temperature in this ideal way. In the Carnot† cycle (see Fig. 3.1), any working substance receives an amount of heat

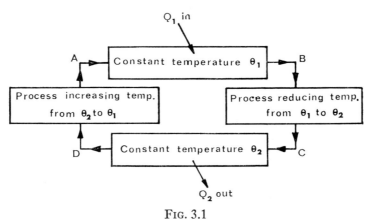

Fig. 3.1

Q_1 while at a temperature θ_1 (this temperature being specified according to any suitable scale, i.e. set of numbers); the substance then undergoes some process (for example, an expansion) such that its temperature is reduced to θ_2 but without any friction or transfer of heat; an amount of heat Q_2 is given up at the temperature θ_2; finally the substance undergoes another process (for example, a compression) by which, again without friction or transfer of heat, the temperature is raised once more to θ_1. During this cycle a net amount of mechanical work W is performed by the substance: thus the efficiency of the engine is

$$\eta = \frac{W}{Q_1} = \frac{Q_1 - Q_2}{Q_1}$$

It is shown in textbooks of thermodynamics that the efficiency of a Carnot engine is independent of the nature of the working substance and is a function only of the temperatures θ_1 and θ_2. Therefore

$$\frac{Q_1}{Q_2} = \frac{1}{1 - \eta} = \phi(\theta_1, \theta_2)$$

† Pronounced *Kar'-no*. The French physicist Nicolas Léonard Sadi Carnot (1796–1832) was one of the founders of modern thermodynamics.

where $\phi(\theta_1, \theta_2)$ means 'an unknown function of θ_1 and θ_2'. Similarly, for a Carnot engine working between two temperatures θ_2 and θ_3

$$\frac{Q_2}{Q_3} = \phi(\theta_2, \theta_3)$$

If these two Carnot engines shared a common heat reservoir at temperature θ_2, the amount of heat Q_2 rejected by the first engine would be absorbed by the second. Thus the two engines together would constitute a third Carnot engine receiving an amount of heat Q_1 at temperature θ_1 and rejecting an amount Q_3 at temperature θ_3. Consequently

$$\frac{Q_1}{Q_3} = \phi(\theta_1, \theta_3)$$

Since $(Q_1/Q_3) \div (Q_2/Q_3) = Q_1/Q_2$, we have

$$\frac{\phi(\theta_1, \theta_3)}{\phi(\theta_2, \theta_3)} = \frac{Q_1}{Q_2} = \phi(\theta_1, \theta_2)$$

The temperature θ_3 does not appear on the right-hand side of this equation: therefore it must cancel in the left-hand side. Once the cancelling has been done we may write

$$\frac{f(\theta_1)}{f(\theta_2)} = \frac{Q_1}{Q_2} = \phi(\theta_1, \theta_2)$$

where $f(\ )$ represents another unknown function.

If the function $f(\ )$ is specified, then this equation serves to define the set of numbers θ, that is, a scale based on the efficiency of a Carnot engine and thus independent of the nature of the working substance. Because the choice of the function $f(\ )$ is unlimited, an infinite number of 'scales' is possible. However, Lord Kelvin finally chose to set $f(\theta)$ directly proportional to a temperature T so that Q_1/Q_2 is simply the ratio of two numbers expressed by the 'T' scale for temperature. This choice of 'T' scale has been universally accepted.

The efficiency of the Carnot engine may now be written

$$\eta = 1 - \frac{Q_2}{Q_1} = 1 - \frac{T_2}{T_1}$$

Thus, for a given value of T_1, the efficiency increases as T_2, the temperature at which the engine rejects heat, is reduced.

Since a Carnot engine is an ideal impossible to realize in practice, the calibration of thermometers by the measurement of the efficiency η is out of the question. Use may be made, however, of the properties of a 'thermally perfect gas'. Now this may seem like reverting to the properties of a particular substance for defining a 'scale' of temperature. A thermally perfect gas, however, is not a particular substance but a hypothetical one whose behaviour is entirely defined by algebraic equations. Such a gas obeys Boyle's Law, that is, at a fixed temperature θ the product of the absolute pressure p and the volume v per unit mass is constant:

$$pv = f(\theta)$$

Evidently a temperature 'scale' may be constructed by using such a gas in a constant-volume thermometer because, if v is held constant, the temperature θ

is uniquely indicated by p. It is convenient to have a linear relation between p and θ, and so we may set $f(\theta) = R(\theta + \theta_0)$ where R and θ_0 are constants. As we shall see later, if the melting point of ice and the boiling point of pure water at 1 atmosphere pressure are respectively termed $\theta = 0°$ and $\theta = 100°$, then $\theta_0 = 273 \cdot 15°$. Thus, from the hypotheses of a thermally perfect gas and a linear relation between p and θ in a constant-volume thermometer, we have

$$pv = R(\theta + \theta_0) \qquad (3.1)$$

This equation must now be related to the result obtained from considering the Carnot cycle. Textbooks of thermodynamics tell us that the First Law of Thermodynamics may be expressed in the form

$$\delta q_{\text{rev}} = \delta e + p \, \delta v$$

where δq_{rev} represents the heat given to unit mass of the gas during a small part of a reversible (i.e. frictionless) process, δe represents the corresponding increase in internal energy of unit mass of the gas, and δv the corresponding increase in the volume of unit mass. For a thermally perfect gas, e is a function of temperature only. The Carnot cycle, we recall, consists of four parts:

A to B Heat Q_1 is given to working substance at constant temperature θ_1;
B to C The temperature drops to θ_2 without transfer of heat to or from the working substance;
C to D Heat Q_2 is rejected by working substance at constant temperature θ_2;
D to A The temperature is increased again to θ_1 without transfer of heat to or from the working substance.

If, as in the second part of the Carnot cycle, no heat is transferred to or from the gas, $\delta q_{\text{rev}} = 0$, whence

$$\delta e + p \, \delta v = 0$$

and division by $\theta + \theta_0$ gives

$$-\frac{\delta e}{\theta + \theta_0} = \frac{p \, \delta v}{\theta + \theta_0}$$

So, for the entire change from B to C,

$$\int_{\theta_1}^{\theta_2} -\frac{de}{\theta + \theta_0} = \int_B^C \frac{p \, dv}{\theta + \theta_0} = \int_B^C \frac{R \, dv}{v} = R \ln\left(\frac{v_C}{v_B}\right) \qquad (3.2)$$

in which we have made use of eqn 3.1.

Similarly, for the fourth part of the cycle (from D to A),

$$\int_{\theta_2}^{\theta_1} -\frac{de}{\theta + \theta_0} = R \ln\left(\frac{v_A}{v_D}\right) \qquad (3.3)$$

Since the left-hand sides of the expressions 3.2 and 3.3 are equal but opposite, $v_C/v_B = v_D/v_A$ and so

$$\frac{v_C}{v_D} = \frac{v_B}{v_A} \qquad (3.4)$$

Since, for the thermally perfect gas considered, e depends on temperature only, $\delta e = 0$ during the constant-temperature process A to B, and so

$$Q_1 = \int_A^B dq_{\text{rev}} = \int_A^B p \, dv$$

Again using eqn 3.1, we have

$$Q_1 = \int_A^B p\, dv = \int_A^B \frac{R(\theta_1 + \theta_0)}{v}\, dv = R(\theta_1 + \theta_0) \ln\left(\frac{v_B}{v_A}\right) \quad (3.5)$$

A similar expression applies to the process from C to D, but Q_2 is negative because heat is then given *out* by the gas:

$$Q_2 = -R(\theta_2 + \theta_0) \ln\left(\frac{v_D}{v_C}\right) \quad (3.6)$$

From eqns 3.5 and 3.6, and then 3.4, we obtain

$$\frac{Q_1}{Q_2} = \frac{(\theta_1 + \theta_0) \ln (v_B/v_A)}{(\theta_2 + \theta_0) \ln (v_C/v_D)} = \frac{\theta_1 + \theta_0}{\theta_2 + \theta_0}$$

If we compare this result with Lord Kelvin's definition by which $Q_1/Q_2 = T_1/T_2$ we have

$$\frac{\theta_1 + \theta_0}{\theta_2 + \theta_0} = \frac{T_1}{T_2}$$

Thus $T = k(\theta + \theta_0)$, and by arbitrarily setting the constant k equal to unity the size of 'degrees' on the 'T' scale is made the same as the size of those on the 'θ' scale. If the 'θ' scale has 100 degrees between the 'ice point' and the 'steam point', then the corresponding 'T' scale is known as the Kelvin absolute scale. The units on the scale used to be termed 'degrees Kelvin', abbreviated °K. Modern practice, however, is to name the unit simply *kelvin*, abbreviated 'K'.

For the relation between T and θ to be completely defined, θ_0 must be known. If the size of degree is already specified (as by defining 100 degrees as the difference in temperature between the 'ice point' and the 'steam point') then θ_0 must be determined by measurement. It might be thought that the value of θ_0 thus obtained would depend on the nature of the gas used in the constant-volume thermometer. However, as the amount, and therefore the pressure, of gas in the fixed volume is reduced, results for θ_0 tend to the same value for all gases (provided that they are not close to liquefaction or dissociation). This value is 273·15 K. (In practice, measurements are usually made with helium, chiefly because of its chemical stability and its very low boiling point. To account for small departures from the equation $pv = R(\theta + \theta_0)$ corrections to the results may be made by a method of successive approximations.) Once the Kelvin scale has been completely defined, one can calibrate ordinary mercury-in-glass thermometers, for example, on which each degree interval is the same size as one on the Kelvin absolute scale (i.e. 1 kelvin), regardless of the coefficients of expansion of the mercury or the glass.

Since 1954 the melting point of ice has been abandoned as a fixed temperature because it is difficult to reproduce it with sufficient accuracy. The condition required is equilibrium between pure ice and air-saturated water. The melting of the ice, however, produces pure water which tends to insulate the ice from the air-saturated water. Consequently the value 273·15 K has an uncertainty of about 0·02 K, and at temperatures approaching the 'absolute zero' ($T = 0$) the resulting lack of precision in values of absolute temperature (that is, temperature expressed by the 'T' scale) becomes serious.

A more satisfactory fixed point is the 'triple point' of water substance, that is the temperature at which ice, liquid water and water vapour exist together in equilibrium at a pressure of 1 standard atmosphere ($1 \cdot 013\,25 \times 10^5$ N/m²). This temperature is reproducible to an accuracy of about $0 \cdot 001$ K and is about $0 \cdot 01$ K above the 'ice point'. By international agreement the value $273 \cdot 16$ K exactly has been adopted for the triple point. Other 'absolute' temperatures can then be determined by a constant-volume gas thermometer:

$$\frac{T}{273 \cdot 16°} = \frac{\text{Pressure at temperature } T}{\text{Pressure at triple point}} \quad (v \text{ constant})$$

3.4.2 Practical Scales of Temperature

3.4.2.1 The Celsius Scale

The Celsius scale (formerly called the centigrade scale) was originally defined by assigning the value 0°C to the melting point of ice and 100°C to the boiling point of water at standard atmospheric pressure. However, when the triple point of water substance was selected as a standard temperature the Celsius scale was redefined by fixing its zero at $273 \cdot 15$ K exactly. Thus a temperature of t°C corresponds to $(273 \cdot 15 + t)$ K.

(Until recently it was usual to distinguish between a particular temperature, that is a definite point on the scale, and a temperature difference. This was done by writing, for example, 50°C for the point on the scale but 50 degC for the difference between, say, 25°C and 75°C. However, now that temperatures are specified without reference to the properties of a particular substance, the definition of the unit of temperature is as fundamental and precise as the definition of units of length or mass; thus there is no virtue in preserving this distinction. The unit of temperature is now termed simply the *kelvin*, abbreviated 'K', and this is used both for the 'absolute' temperature and for differences of temperature.)

3.4.2.2 The Fahrenheit Scale

This scale, which has had wide use in Great Britain and North America, was formerly defined by two fixed points: 32°F for the 'ice point' and 212°F for the 'steam point'. The range of 180 degrees between these two points corresponded to 100 degrees on the Celsius scale, and so a Fahrenheit degree is five-ninths of a Celsius degree. A temperature on one scale may be expressed as one on the other scale by the formulae

$$C = \tfrac{5}{9}(F - 32) \quad \text{and} \quad F = \tfrac{9}{5}C + 32$$

where C represents the number of degrees on the Celsius scale, and F the number of degrees on the Fahrenheit scale at the same temperature.

Absolute zero temperature on the Fahrenheit scale is at $\tfrac{9}{5}(-273 \cdot 15) + 32 = -459 \cdot 67°$. By adding $459 \cdot 67°$ to a temperature expressed on the Fahrenheit scale we therefore obtain an 'absolute temperature' in terms of degrees Rankine (abbreviated °R); that is, a temperature of t°F corresponds to $(459 \cdot 67 + t)$°R.

3.4.2.3 The International Practical Temperature (IPT) Scale

Direct calibration of thermometers and similar devices against a constant-volume gas thermometer is difficult and time-consuming. A practical scale has therefore been internationally agreed upon, not to supplant the Kelvin thermodynamic definition of the magnitude of temperature, but to provide a set of values for various fixed points which can readily be used for calibrations. Some of these values, given in the 1968 revision of the scale, are listed in Table 3.1. All are for a pressure of 1 standard atmosphere.

TABLE 3.1

	°C	K
Boiling point of hydrogen (the 'hydrogen point')	−252·87	20·28
Boiling point of oxygen (the 'oxygen point')	−182·962	90·188
Equilibrium of ice and air-saturated water (the 'ice point')	0·00	273·15
Boiling point of water (the 'steam point')	100·00	373·15
Melting point of zinc (the 'zinc point')	419·58	692·73
Melting point of silver (the 'silver point')	961·93	1235·08
Melting point of gold (the 'gold point')	1064·43	1337·58

Methods of calibration to be employed between the oxygen point and the gold point have been agreed on. For higher temperatures an optical method is used based on Planck's radiation law. In the measurement of very low temperatures use is made of magnetic properties.

The accuracy of this practical temperature scale depends of course on the accuracy of the individual measurements on which it is based. Indeed, modifications of some of the values may prove necessary as more reliable experimental data become available. For most calculations, however, errors in the IPT scale are negligible.

3.4.3 UNITS FOR QUANTITY OF HEAT

When a hot body is brought into contact with a cooler body, so that the temperatures of the two bodies tend to equalize, heat is said to pass from the hotter body to the cooler. Heat is now recognized as a form of energy, and so may be measured in terms of units of energy. In fact, since 1948, the basic unit of quantity of heat has been that of mechanical energy, the *joule* (i.e. metre newton), and the magnitudes of all other units for quantity of heat have been defined in terms of the joule.

However, many units for quantity of heat have been separately defined, and some of these are still in wide use. Each unit has been defined as the amount of heat required to raise the temperature of a standard amount of a standard substance from one standard temperature to another without change of pressure or of phase (e.g. from liquid to gas). The standard substance has invariably been pure water, but various values have been used for the standard amount of it and particularly for the standard temperature difference.

The *calorie* (abbreviated 'cal') is defined as the amount of heat required to raise the temperature of 1 gram mass of pure water by 1 degree Celsius at 1 atmosphere pressure. However, this simple definition is inexact because the specific heat capacity of water varies somewhat with temperature. Unfortunately, there was no universal agreement about where on the temperature scale the one-degree interval should be taken. As a result, a number of slightly different calories came into being. Among the definitions were temperature ranges from 0°C to 1°C, from 3·5°C to 4·5°C, from 19·5°C to 20·5°C, and from 14·5°C to 15·5°C. This last, the most usual in scientific work, is known as the '15° calorie'. Another definition specified the calorie as one-hundredth of the quantity of heat required to raise 1 gram mass of pure water from 0°C to 100°C. There was a range of about 0·7 per cent in the size of these various calories.

When the unit of heat quantity is defined without reference to other forms of energy, the quantitative relation between heat and other forms of energy is a matter for experimental determination. Since the first measurements by J. P. Joule† in 1840, many results for the 'mechanical equivalent of heat' have been obtained and the most accurate indicate that one 15° calorie is equivalent to 4·1858 joules (where 1 joule ≡ 1 metre newton).

Yet another sort of calorie in use was the *international steam table calorie* (IT cal), originally defined in terms of electrical units but later fixed at 4·1868 joules exactly. (The calorie used as a unit of food value is actually a kilocalorie, that is 1000 times the 15° calorie.)

The *British thermal unit* (abbreviated Btu), like the calorie, is defined in terms of a standard amount of pure water and a standard temperature interval. The standard amount of water is 1 lbm (at a constant pressure of 1 atmosphere) and the temperature interval has frequently been from 60°F to 61°F. Other intervals have also been used. The mean Btu is defined as 1/180 times the quantity of heat required to raise the temperature of 1 lbm of pure water from 32°F to 212°F. The Btu used in steam tables and in similar data is based on the IT calorie. A heat unit, used particularly in the gas industry, is the *therm*, defined as 10^5 Btu.

The *Celsius heat unit*‡ (abbreviated CHU) may be defined (approximately) as the quantity of heat required to raise the temperature of 1 lbm of pure water by one degree Celsius at a constant pressure of 1 atmosphere. The precise definition is 1 CHU = 453·592 37 IT cal.

The relations connecting the magnitudes of the various units of heat quantity are given in Chapter 5.

3.5 Units for Electric and Magnetic Quantities

The magnitudes of quantities involved in electric and magnetic phenomena cannot all be defined entirely in terms of units of length, mass and time interval. Some additional base unit is needed. The choice of this additional unit, like the choice of the other three, is arbitrary, and, because different choices have been made, different systems of units have been developed. Moreover, some of the derived units have been defined in terms of the base units by equations in which the arbitrary numerical constants have had alternative values.

Although it is not the purpose of this book to record the tangled history of electric and magnetic units, some reference will be made to units which are no longer standard. These are mentioned because they still have appreciable (though decreasing) use and they are found in many published data to which reference is still often made. No attempt will be made, however, to mention all units which have been proposed.

† Pronounced *Jool*. ‡ Formerly termed Centigrade heat unit.

3.5.1 The c.g.s. Electrostatic System of Units (c.g.s. e.s.u.)

This system is based on the centimetre, the gram mass and the second as units of length, mass and time interval respectively, and, as a fourth base unit, a unit of permittivity. This last quantity figures in Coulomb's Inverse Square Law for electrostatic charges

$$F \propto \frac{Q_1 Q_2}{\varepsilon r^2} \tag{3.7}$$

in which F represents the force exerted between two isolated bodies having electric charges Q_1 and Q_2, respectively, and separated by a distance r which is large compared with the size of the bodies. The symbol ε (Greek 'epsilon') represents a factor, the value of which depends on the medium between the bodies, and is termed the *permittivity* of the medium. (It was formerly known as the *specific inductive capacity* or the *dielectric constant*—although strictly it is constant only for very small values of Q_1 and Q_2.) The medium having the smallest value of ε is a vacuum; air and most other gases at ordinary pressures have values very little more, but liquids and solids may have values up to 100 times that of a vacuum. (Occasionally, for a crystalline medium, the value of ε may depend on the direction of the distance r in relation to the arrangement of the crystals.)

If the constant of proportionality in the expression 3.7 is arbitrarily put at unity, and if the permittivity of a vacuum is arbitrarily taken as the unit of permittivity, then the equation serves to define the unit of electric charge Q. Since the unit of force in the c.g.s. system is the dyne ($=$ g cm s^{-2}), the unit of charge is that of $(F\varepsilon r^2)^{1/2}$, i.e.

$$\text{cm}^{3/2} \text{ g}^{1/2} \text{ s}^{-1} \text{ (unit of } \varepsilon)^{1/2}.$$

Current, defined as rate of transfer of charge, then has as its unit

$$\text{cm}^{3/2} \text{ g}^{1/2} \text{ s}^{-2} \text{ (unit of } \varepsilon)^{1/2}.$$

Unit difference of electric potential is defined as that which would allow unit charge to do 1 unit of work (i.e. 1 cm dyne $=$ 1 cm^2 g s^{-2}) and so

$$\text{unit of potential difference} = \frac{\text{unit of work}}{\text{unit of charge}} = \frac{\text{cm}^2 \text{ g s}^{-2}}{\text{cm}^{3/2} \text{ g}^{1/2} \text{ s}^{-1} \text{ (unit of } \varepsilon)^{1/2}}$$

$$= \text{cm}^{1/2} \text{ g}^{1/2} \text{ s}^{-1} \text{ (unit of } \varepsilon)^{-1/2}.$$

Units of other electric quantities may be derived from these.

It is seldom that the units are given a distinctive name: they are usually referred to simply as electrostatic units of current, potential difference and so on. However, the e.s.u. of charge has occasionally been termed the *franklin* (Fr). Also, the prefix 'stat' has sometimes been used with the name of the modern unit (see Section 3.5.4) for the same quantity: for example, a 'statampere' is the e.s.u. of current, and a 'statfarad' the e.s.u. of capacitance. This terminology, however, should not blind us to the fact that the sizes of the electrostatic units differ—often considerably—from those of the corresponding modern units.

3.5.2 The c.g.s. Electromagnetic System of Units (c.g.s. e.m.u.)

Again the centimetre, gram mass and second are used as three of the base units, but in this system the fourth is the unit of magnetic permeability, this being chosen as equal to the permeability of a vacuum.

The magnetic permeability μ arises in the expression for the magnitude of the force δF between two elements of current-carrying conductors

$$\delta F \propto \frac{\mu(I_1 \, \delta l_1 \sin \theta_1)(I_2 \, \delta l_2 \sin \theta_2)}{r^2} \tag{3.8}$$

where currents I_1 and I_2 flow in elements of length δl_1 and δl_2, respectively, which are separated by a distance r. The angles θ_1, θ_2 are those between the direction of the distance r and each of the conductors. The value of μ depends on the medium between the conductors (and occasionally on the direction of the distance r relative to the arrangement of crystals in a crystalline medium).

If the coefficient of proportionality in the expression 3.8 is arbitrarily set at unity, then the resulting equation defines the unit of current as that of

$$\left(\frac{\delta F \, r^2}{\mu \, \delta l_1 \, \delta l_2} \right)^{1/2}$$

i.e.

$$\left\{ \frac{\text{dyne cm}^2}{(\text{unit of } \mu) \, \text{cm}^2} \right\}^{1/2} = \text{cm}^{1/2} \, \text{g}^{1/2} \, \text{s}^{-1} \, (\text{unit of } \mu)^{-1/2}$$

(This unit was sometimes termed *abampere* and occasionally *biot*.†)

The corresponding unit of charge is therefore $\text{cm}^{1/2} \, \text{g}^{1/2} \, (\text{unit of } \mu)^{-1/2}$ (sometimes termed *abcoulomb*). A unit of potential difference is then deducible as

$$\frac{\text{unit of work}}{\text{unit of charge}} = \frac{\text{cm dyne}}{\text{cm}^{1/2} \, \text{g}^{1/2} \, (\text{unit of } \mu)^{-1/2}} = \text{cm}^{3/2} \, \text{g}^{1/2} \, \text{s}^{-2} \, (\text{unit of } \mu)^{1/2}$$

From these units, units of other electric and magnetic quantities may be determined from the definitions of their magnitudes. A few units in the c.g.s. electromagnetic system have been given distinctive names, and these will be found at the appropriate places in Appendix 1. Here we need to draw attention only to the names of the electromagnetic units for magnetic flux density, B, and magnetic field strength, H. The magnitudes of these quantities are connected by the relation $B = \mu H$. In the e.m.u. system, μ for a vacuum is 1 unit of μ, and, under normal conditions, the value for air and other gases is negligibly different. Hence, with these units, values of B and H for a vacuum or a gaseous medium are *numerically* (though certainly not physically) the same. As a result, the named unit for B, the gauss‡, was frequently applied also to values of H. In 1930 the unit of H was named 'oersted'§, leaving the name 'gauss' to be used only for the unit of B. Bad habits, however, are difficult to eradicate, and the widespread misuse of the name 'gauss' continued for many years.

In the late nineteenth century the name 'gauss' was used occasionally for units of various magnetic quantities. Therefore, whenever a unit named 'gauss' is encountered it is always necessary to examine the context carefully to discover to which quantity the unit applies.

Sometimes units of the e.m.u. system have been named by using the prefix 'ab' with the name of the modern unit (see Section 3.5.4) for the same quantity. For example, 'abohm' has been used for the e.m.u. of electrical resistance, and 'abvolt' for the e.m.u. of potential difference. This terminology, however, should

---

† Pronounced *be'-oh*.   ‡ Pronounced *gowce*.   § Pronounced *er'-sted*.

§3.5]   UNITS FOR ELECTRIC AND MAGNETIC QUANTITIES   33

not be allowed to obscure the fact that e.m. units differ in size from the corresponding modern units.

It is shown in text-books of electromagnetic theory that the magnitude of the velocity c with which electromagnetic waves (such as those of light) are propagated is given by

$$c = \frac{1}{(\varepsilon\mu)^{1/2}}$$

where ε represents the permittivity of the medium (strictly, in the direction of c) and μ its magnetic permeability (in the same direction). Since the velocity of such waves in a vacuum is $2{\cdot}997\,93 \times 10^{10}$ cm/s, the values of ε and μ for a vacuum cannot simultaneously each be 1 unit when cm/s is the unit of velocity. Consequently, with few exceptions, electrostatic units and electromagnetic units for particular quantities differ greatly in size. For example, the ratio

$$\frac{1 \text{ c.g.s. electrostatic unit of potential difference}}{1 \text{ c.g.s. electromagnetic unit of potential difference}}$$

$$= \frac{1 \text{ cm}^{1/2}\,\text{g}^{1/2}\,\text{s}^{-1}\,(\text{unit of } \varepsilon)^{-1/2}}{1 \text{ cm}^{3/2}\,\text{g}^{1/2}\,\text{s}^{-2}\,(\text{unit of } \mu)^{1/2}} = \frac{\text{s}}{\text{cm}\,(\text{unit of } \varepsilon)^{1/2}\,(\text{unit of } \mu)^{1/2}}$$

Since $c^2 = (\varepsilon\mu)^{-1}$, $\mu = (c^2\varepsilon)^{-1}$ and

the unit of $\mu = (2{\cdot}997\,93 \times 10^{10} \text{ cm s}^{-1})^{-2}\,(\text{unit of } \varepsilon)^{-1}$.

Therefore the ratio

$$\frac{1 \text{ e.s.u. of potential difference}}{1 \text{ e.m.u. of potential difference}}$$

$$= \frac{\text{s}}{\text{cm}\,(\text{unit of } \varepsilon)^{1/2}}\left(2{\cdot}997\,93 \times 10^{10}\,\frac{\text{cm}}{\text{s}}\right)(\text{unit of } \varepsilon)^{1/2}$$

$$= 2{\cdot}997\,93 \times 10^{10}.$$

(Particularly in atomic and nuclear physics, a combination of the electrostatic and electromagnetic systems of units has sometimes been used. Usually associated with this simultaneous use of units from both systems was the treatment of algebraic equations as 'ordinary', not 'physical' algebra—that is, the equations were regarded only as relations connecting numbers. To make the equations suitable for use with the combined set of units, it was then necessary to adjust some of the equations by inserting into them, as a factor, some power of $2{\cdot}997\,93 \times 10^{10}$. This whole procedure—which modern thought is bound to regard as unenlightened—was curiously termed 'using the Gaussian system of units', although the units themselves were the same as those of the e.s. and e.m. systems.)

3.5.3 THE PRACTICAL SYSTEM OF UNITS

Since neither the c.g.s. electrostatic nor the c.g.s. electromagnetic units are of convenient size for engineering use, a practical system of units grew up based on a unit of electric current (the ampere) and a unit of potential difference (the volt). These units were by definition related to the c.g.s. electromagnetic units by integral powers of 10: the ampere was defined as 10^{-1} e.m.u. of current and the volt as 10^8 e.m.u. of potential difference. As the need for other units arose they were defined in terms of the centimetre, gram mass, second, ampere and volt, and therefore these units also were related to corresponding e.m. units by integral powers of 10.

This practical system was sufficient for most ordinary engineering calculations but was in some respects incomplete. To fill the gaps in it engineers used units from the c.g.s. electromagnetic system, and therefore had to use conversion factors (see Chapter 5) in their calculations.

To fix the size of units exactly for calibration purposes, precisely reproducible standards and accurate methods of direct comparison are necessary. There is no reason why the standards against which calibrations are made should be those of quantities used in basic definitions. The exactly reproducible standards may be those of any suitable quantities, and the magnitude of the units of the fundamental quantities can be deduced by 'working backwards'.

Accurate determinations of electrical units may be made, for example, in terms of the unit of current as measured by elaborate 'current balances'. However, such measurements are difficult to make and so a number of 'International' standards were defined, which, while representing to a high degree of accuracy units already defined as multiples of electromagnetic or electrostatic units, could be readily set up by standards laboratories in various countries. Thus, by international agreement in 1908, the standard unit of resistance, the ohm, was defined as the resistance of a certain column of mercury. At the same time a standard unit of electric current, the ampere, was defined in terms of the rate of deposition of silver from a standard solution of silver nitrate through which the current passed. These units, and others derived from them, were termed 'International Units'.

However, later and more exact measurements showed that the international ohm and ampere differed slightly from those defined as multiples of electromagnetic units. This discrepancy provided an additional reason for changing to a single system for all electric and magnetic units.

3.5.4 THE INTERNATIONAL MKSA SYSTEM OF UNITS

The present system of electric and magnetic units uses the metre, the kilogram mass and the second as the three mechanical base units, and the ampere as the fourth base unit. The more commonly used electrical units in this system are the same (to within close limits) as those in the earlier 'practical' system (Section 3.5.3) but they are given more exact definitions. By international agreement the *ampere* is now defined as that constant current which, if maintained in each of two straight parallel conductors, of infinite length and negligible cross-section, placed one metre apart in a vacuum, produces on each conductor a force equal to 2×10^{-7} newtons per metre length.

From this unit all other electric and magnetic units may be derived. A steady current of 1 ampere (A) passing for 1 second carries 1 unit of charge (i.e. quantity of electricity) which is termed 1 *coulomb* (C).

The unit of work or energy, the *joule* (J), equal to metre newton, is derived from the mechanical units. If an amount of work W is done on a small charge of magnitude Q in order to move it from point A to point B, then the limit, as $Q \to 0$, of the quotient W/Q is defined as the difference of electric potential between the points A and B. Thus the unit of potential difference, the *volt* (V) is defined as 1 joule divided by 1 coulomb. The unit of power is the joule per second, termed the *watt* (W). A steady current of 1 ampere carries charge at the rate of 1 coulomb per second: thus such a current flowing through a conductor will dissipate 1 watt between two points with a potential difference of 1 volt.

The *ohm* (Ω), the unit of electrical resistance, is the resistance between two points in a conductor when a constant potential difference of 1 volt, applied between these two points, produces a current of 1 ampere (provided that the conductor is not the source of any electromotive force).

The *farad* (F), the unit of electric capacitance, is the capacitance of a capacitor which acquires a potential difference of 1 volt across its plates when it is charged with 1 coulomb of electricity.

The *weber*[†] (Wb) is that magnetic flux which, linking a circuit consisting of one complete turn of wire, produces in it an e.m.f. of 1 volt as the flux is reduced to zero at a uniform rate in 1 second.

The *henry* (H) is the magnitude of the inductance of a closed circuit in which an e.m.f. of 1 volt is produced when the electric current in the circuit varies uniformly at the rate of 1 ampere per second.

The *tesla* (T) is a unit of magnetic flux density equal to 1 weber per square metre of circuit area.

3.5.5 Rationalized Equations

In the c.g.s. electrostatic and c.g.s. electromagnetic systems of units, the coefficients of proportionality in the expressions 3.7 and 3.8 were arbitrarily set equal to unity. A consequence of using the resulting equations to define the magnitudes of electric charge or current was that the numerical factor 4π appeared in other relations which were concerned only with rectilinear geometrical arrangements, yet 4π did not appear in equations concerned with spherical geometrical arrangements. (An analogous state of affairs would have arisen if unit area had been defined, not as the area of a square with sides of unit length, but as the area of a circle of unit radius. The area of a circle—which, as experiment shows, is proportional to the square of the radius r—would then have been given by $A = r^2$ instead of $A = \pi r^2$, and the area of a rectangle with sides of lengths a and b would have been given by ab/π. Since π is associated by definition with a circle, it is illogical that it should appear in the formula for the area of a rectangle, yet not in the formula for the area of a circle.)

This illogicality in equations describing electric and magnetic phenomena was removed by setting the proportionality constants in the expressions 3.7 and 3.8 equal to $(1/4\pi)$ rather than unity. However, since in the electrostatic or electromagnetic systems of units one of these expressions was used to define unit charge (or current) and thence all other electrical units, the unit of permittivity ε or permeability μ was made 4π times larger than originally. Thus the sizes of the more commonly used units were preserved unchanged.

The introduction of $1/(4\pi)$ in the expressions 3.7 and 3.8 and consequent alterations in some other equations was termed 'rationalizing' the equations. The 'rationalized' equations are those found in all modern electrical text-books. Moreover, the present internationally standardized units discussed in Section 3.5.4 are all based on 'rationalized' equations, and so the former 'unrationalized'

[†] Pronounced *Vay'ber*.

equations, like the c.g.s. electrostatic and the c.g.s. electromagnetic systems of units, will soon be of little more than historical interest.

3.6 Units for Quantities connected with Illumination

An additional base unit is required to express the magnitudes of quantities involving light. This unit has always been one of luminous intensity, that is the intensity of a source of light. Originally the luminous intensity of a 'standard candle' was used, and thus luminous intensity was often colloquially termed 'candle-power'. However, accurate reproduction of the standard candle was difficult, and the unit was superseded by one defined in terms of the intensity of light from a lamp of standard construction in which a standard gas was burnt. This in turn has now been superseded by the *candela*† (abbreviated 'cd'). In magnitude the candela corresponds closely to the intensity of the former standard candle, but it can be much more exactly defined:

The candela is the intensity of light emitted in a perpendicular direction from an area of $1/600\,000$ m² of a black-body (i.e. a perfect) radiating surface at the freezing point of pure platinum (1773°C) under a pressure of 1 standard atmosphere (101 325 N/m²).

The unit of luminous flux (that is, the rate of flow of light energy) is the *lumen* (lm). If a 'standard candle' (or, more precisely, a point source of uniform intensity 1 candela) were placed at the apex of a cone of unit solid angle (1 steradian), then the light energy steadily streaming outwards in the cone would be 1 lumen. A lumen is thus the rate at which light energy falls perpendicularly on unit area of a spherical surface, of unit radius, at the centre of which is a uniform point source of intensity 1 candela. (From an ordinary 100 W tungsten-filament electric light bulb the total luminous flux in all directions is about 1700 lumens.)

The degree of illumination of a surface on which light falls is termed the luminous flux density, or illumination, and is expressed as the rate at which light energy arrives at unit area of the surface. An appropriate unit is therefore lumen/metre², also termed *lux* (lx). It corresponds to the amount of light arriving at unit area of a surface which directly faces a standard candle 1 metre away from it. For this reason the lux has sometimes been termed the *metre-candle*. Another unit still widely used is the foot-candle (f.c.): it corresponds to the illumination arriving perpendicularly at a surface facing a standard candle 1 foot away, and so equals lm/ft².

The luminance of a surface is a measure of its brightness, that is, of the light emitted or reflected from it in a particular direction, as distinct from incident light. Luminance, in a specified direction, is therefore expressed as the ratio of the luminous intensity of a small element of the area (observed from the given direction to the orthogonal projection of the area element in the plane perpendicular to the direction of observation). An appropriate unit is thus candela/metre². However, the luminance of a surface is sometimes expressed, not in terms of its luminous intensity per unit area, but in terms of the luminance of a perfectly diffusing surface emitting or reflecting 1 lumen/metre². Most matt surfaces appear equally bright from all directions, that is, they diffuse uniformly the

---

† Pronounced *kan-dell'-ar*: the word is the Latin for 'candle'.

light they reflect. The luminance refers only to the light emitted or reflected perpendicularly to the surface, whereas it may be shown that for a uniformly diffusing surface the *total* flux emitted by unit area is equal to the luminance multiplied by π. This has given rise to a unit known as the *apostilb* (asb): a uniformly diffusing surface with a luminance L cd/m^2 emits πL apostilb. The corresponding unit based on an area of 1 cm^2 is known as the *lambert* (L), and that based on an area of 1 ft^2 is, rather confusingly, termed the *foot-lambert* (ft-L).

For any surface other than a uniformly diffusing one, however, the luminance in a particular direction is not definitely related to the total flux emitted from the surface but only to the luminous intensity seen from that direction.

Photometry, that is the measurement of light, is a branch of science in which a wide variation of terminology may be found and many units have been devised other than those mentioned here.

4

The Système International d'Unités

4.1 Introduction

As science and engineering have developed, units and relations among units have been devised in great variety. The reader may feel that those mentioned in this book are bewildering in their profusion, yet only the most important units have been included here. Many units which have been suggested for expressing the magnitudes of particular quantities have had limited use and have then been forgotten. Even so, we have seen that some quantities, such as force, are still expressed in terms of any of several units all in wide use, and thus it is often necessary to convert a magnitude expressed in terms of one unit into an alternative form involving a unit of the same kind but of different size. Not only does this call for much arithmetic which would otherwise be unnecessary, but the possibility of confusion and error is greatly increased.

For many years, dissatisfaction with this state of affairs has been growing, and the advantages of a single, internationally agreed system of units have been increasingly realized. In 1960, the General Conference of Weights and Measures (an international body founded in 1875 to promote uniformity in standards of measurement) finally agreed that the system to be recommended for general adoption should be a version of the 'metric system' known as the Système International d'Unités—abbreviated 'S.I.'. This system is being adopted throughout much of the world and there seems no doubt that those countries

TABLE 4.1

Quantity	Unit	Symbol
length	metre	m
mass	kilogram†	kg†
time interval	second	s
electric current	ampere	A
temperature	kelvin	K
	(formerly 'degree Kelvin')	(formerly °K)
luminous intensity	candela	cd

† It is probable that the name and the symbol (but not the definition) of this unit will be changed in the future. The forms given here, however, are those internationally standardized at present.

§4.2]   PREFIXES AND CONVENTIONS IN PRINTING   39

which have not yet decided to change to S.I. (principally the U.S.A. and Canada) will later do so.

The names of units in S.I. and the standard symbols for the names have already been mentioned in reference to the various kinds of physical quantity. However, it is useful to gather the names and symbols together here. The six base units are listed in Table 4.1. (A quantity which has been suggested for addition to this list is 'amount of substance', with the unit 'mole'—see Section 3.2.4.)

Certain combinations of the base units shown in Table 4.1 are given internationally agreed special names, see Table 4.2.

TABLE 4.2

Quantity	Unit	Symbol	Equivalent combination of other units
plane angle	radian	rad	m/m (i.e. length of circular arc ÷ radius)
solid angle	steradian	sr	m^2/m^2 (i.e. area of spherical surface ÷ $radius^2$)
force	newton	N	$kg\,m/s^2$
work, energy, quantity of heat (but *not* torque)	joule (pronounced *jool*)	J	Nm
power	watt	W	J/s
electric charge	coulomb	C	As
electric potential	volt	V	J/C
electric capacitance	farad	F	C/V
electric resistance	ohm	Ω	V/A
frequency	hertz	Hz	cycle/s
magnetic flux	weber (pronounced *vay'-ber*)	Wb	V s
magnetic flux density	tesla	T	Wb/m^2
inductance	henry	H	V s/A = Wb/A
luminous flux	lumen	lm	cd sr
illumination	lux	lx	lm/m^2

Some other special names have been proposed and may be adopted in the future.

4.2 Prefixes and Conventions in Printing

If only the base units and their combinations were used in expressing the magnitudes of physical quantities, there would be many occasions when the associated numerics would be inconveniently large or small. To avoid this difficulty, a prefix may be put in front of any unit name. This prefix unites with the name of the base unit or combination of units to give the name of another unit (for the same quantity) greater or smaller than the original unit by a specified numerical factor. For example, the prefix 'kilo' means one thousand. Combining this with 'metre' gives 'kilometre', a unit of length which is one thousand times the metre. Any prefix is always the same, whatever the unit name to which it is applied.

The full list of prefixes is as follows.

Prefix	Symbol	Numerical factor by which basic unit is multiplied
tera	T	10^{12}
giga (pronounced *jy'-ga*)	G	10^9
mega (pronounced *meg'-ga*)	M	10^6
kilo	k	10^3
hecto	h	10^2
deca	da	10
deci	d	10^{-1}
centi	c	10^{-2}
milli	m	10^{-3}
micro	μ	10^{-6}
nano	n	10^{-9}
pico (pronounced *pee'-ko*)	p	10^{-12}
femto	f	10^{-15}
atto	a	10^{-18}

Especially recommended are prefixes which refer to factors of 10^{3n}, where n is a positive or negative integer.

Care is needed in using these prefixes. A prefix should always be written close to the unit it qualifies, e.g. kilometre (km), megawatt (MW), microsecond (μs). On the other hand, the symbols for the basic units (that is, those without prefixes) should be spaced apart, e.g. N s/m^2 or kg/s m^2. Only one prefix at a time may be applied to a unit; thus one thousand kilograms is one megagram (Mg), not one kilo-kilogram.

The symbol 'm' stands for the basic unit 'metre' and for the prefix 'milli', and so especial care is needed in using it. For example, mN means millinewton, whereas m N denotes metre newton. In this case the difference of spacing is hardly sufficient safeguard against confusion, and so, if the 'm' denotes 'metre' it is better to reverse the order of the unit symbols: N m.

When a multiple of a basic unit is raised to a power, the exponent applies to the *whole multiple* and not to the basic unit alone. Thus 1 mm^2 means 1 (mm)^2 = $(10^{-3}$ m$)^2$ = 10^{-6} m^2, and *not* 1 m(m^2) = 10^{-3} m^2. (The abbreviations 'sq' for 'square', and 'cu' for 'cubic', are no longer used.)

If a derived unit is in the form of a quotient, any prefix should normally be applied only to the numerator and not to the denominator. For example, MN/m^2, not N/mm^2, should be used for 10^6 N/m^2.

The symbols for units refer not only to the singular but also to the plural. For instance, the symbol for kilometres is km, not kms. Dots, to denote either multiplication or abbreviation, are not used at all.

Capital or lower case (small) letters are used strictly in accordance with the definitions, no matter in what combination the letters may appear. For example, kilonewton metre is represented by kN m, even though a capital letter appearing between two lower case letters may look a little strange. In print, ordinary upright type is used for unit symbols and prefixes, whereas italic type (*thus*) is used for algebraic symbols.

4.3 Units for Pressure and Stress

The magnitude of any pressure or stress is given by the magnitude of a force divided by the magnitude of an area. Thus the appropriate unit derived directly from basic units is N/m^2. This is small for most purposes and so multiples are often used. Many published data are in terms of the 'bar'. This is defined as $10^5 \ N/m^2$ and has been in use for many years, but as it breaks the 10^{3n} rule it is not an S.I. unit. The special name 'pascal', with symbol 'Pa', has been suggested for N/m^2 and may well be internationally adopted in future. For the present, however, N/m^2 is mostly used, prefixes being put before the N if desired.

5

Conversion Factors

It is frequently desirable to change data involving units of one size so as to involve units of the same kind but of a different size. This may be achieved by using 'conversion factors' by which the relation between the sizes of different units of the same kind may be expressed.

By way of example we may consider the identity

$$1 \text{ inch} \equiv 25 \cdot 4 \text{ mm}$$

Notice the use of three lines here ($\equiv$) instead of the two lines of the usual 'equals' sign. We do not say merely that 1 inch equals 25·4 mm or that 1 inch is equivalent to 25·4 mm, but that 1 inch *is* 25·4 mm. At all times and in all places 1 inch and 25·4 mm are precisely the same magnitude.

This identity may be rewritten as

$$1 \equiv \frac{25 \cdot 4 \text{ mm}}{1 \text{ inch}}$$

and it is this ratio *equal to unity* which is the conversion factor. Since unity may be introduced as a factor into any expression without altering the value of the expression, changing units cannot essentially affect any equation in 'physical' algebra. To change the units is equivalent to multiplying by unity. Moreover, as the reciprocal of unity is also unity any conversion factor may be used in reciprocal form when the desired result requires it.

Our simple example may be extended indefinitely:

$$1 \equiv \frac{25 \cdot 4 \text{ mm}}{1 \text{ inch}} \equiv \frac{1 \text{ inch}}{25 \cdot 4 \text{ mm}} \equiv \frac{304 \cdot 8 \text{ mm}}{1 \text{ ft}} \equiv \frac{0 \cdot 4536 \text{ kg}}{1 \text{ lbm}} \equiv \frac{10^{-5} \text{ N}}{1 \text{ dyne}} \equiv \frac{4 \cdot 448 \text{ N}}{1 \text{ lbf}}$$

$$\equiv \frac{4 \cdot 187 \text{ J}}{1 \text{ calorie}} \equiv \frac{1 \text{ volt}}{10^8 \text{ c.g.s. e.m.u. of potential difference}} \equiv \ldots$$

An application of such conversion factors enables us to express, for example, a force of 4·6 lbf in terms of newtons. Set out fully, the successive operations are:

$$4 \cdot 6 \text{ lbf} = 4 \cdot 6 \text{ lbf} \times \frac{32 \cdot 174 \text{ pdl}}{1 \text{ lbf}} = 4 \cdot 6 \times 32 \cdot 174 \text{ pdl} = 4 \cdot 6 \times 32 \cdot 174 \frac{\text{lbm ft}}{\text{s}^2}$$

$$= 4 \cdot 6 \times 32 \cdot 174 \frac{\text{lbm ft}}{\text{s}^2} \times \frac{0 \cdot 4536 \text{ kg}}{1 \text{ lbm}} \times \frac{0 \cdot 3048 \text{ m}}{1 \text{ ft}}$$

$$= 4 \cdot 6 \times 32 \cdot 174 \times 0 \cdot 4536 \times 0 \cdot 3048 \frac{\text{kg m}}{\text{s}^2}$$

$$= 4 \cdot 6 \times 32 \cdot 174 \times 0 \cdot 4536 \times 0 \cdot 3048 \text{ N} = 20 \cdot 46 \text{ N}.$$

Such detail of setting out is not normally necessary, of course, but this pattern of operations, this successive multiplication by unity, is the logical basis of the use of conversion factors and is therefore implicit in all correct conversions.

As another example, consider the conversion of 50 mile/h to a form involving m/s:

$$50\,\frac{\text{mile}}{\text{h}} = 50\,\frac{\text{mile}}{\text{h}} \times \frac{5280\text{ ft}}{1\text{ mile}} \times \frac{0\cdot3048\text{ m}}{1\text{ ft}} \times \frac{1\text{ h}}{3600\text{ s}} = 22\cdot35\text{ m/s}$$

Or the conversion of 12 lbf/in² to a form involving N/m²:

$$12\,\frac{\text{lbf}}{\text{in}^2} = 12\,\frac{\text{lbf}}{\text{in}^2} \times \frac{32\cdot174\text{ lbm ft/s}^2}{1\text{ lbf}} \times \left(\frac{1\text{ in}}{0\cdot0254\text{ m}}\right)^2 \times \frac{0\cdot3048\text{ m}}{1\text{ ft}} \times \frac{0\cdot4536\text{ kg}}{1\text{ lbm}}$$

$$= 12 \times 32\cdot174 \times \frac{0\cdot3048}{0\cdot0254^2} \times 0\cdot4536\,\frac{\text{kg}}{\text{m s}^2}$$

$$= 8\cdot28 \times 10^4\,\frac{\text{kg}}{\text{m s}^2} = 8\cdot28 \times 10^4\text{ N/m}^2 \text{ or } 82\cdot8\text{ kN/m}^2.$$

A number of useful conversion factors will now be listed. They are derived from the following internationally agreed values.

$$1\text{ ft} \equiv 0\cdot304\,8\text{ m exactly}$$
$$1\text{ lbm} \equiv 0\cdot453\,592\,37\text{ kg exactly}$$

Standard value of weight/mass $\equiv 9\cdot806\,65$ N/kg exactly

1 international steam table calorie $\equiv 4\cdot186\,8$ J exactly

$$1\text{ standard atmosphere} \equiv 1\cdot013\,25 \times 10^5\text{ N/m}^2\text{ exactly}$$

For simplicity the numbers in the following conversion factors are rounded off to three or four significant figures only. In a few cases, however, the exact values require only the number of significant figures given and these are printed in bold type.

Length

$$1 \equiv \frac{\mathbf{12}\text{ in}}{\mathbf{1}\text{ ft}} \equiv \frac{\mathbf{25\cdot4}\text{ mm}}{\mathbf{1}\text{ inch}} \equiv \frac{\mathbf{304\cdot8}\text{ mm}}{\mathbf{1}\text{ ft}} \equiv \frac{\mathbf{914\cdot4}\text{ mm}}{\mathbf{1}\text{ yd}} \equiv \frac{\mathbf{5280}\text{ ft}}{\mathbf{1}\text{ mile}} \equiv \frac{1\cdot609\text{ km}}{1\text{ mile}}$$

$$\equiv \frac{\mathbf{1760}\text{ yd}}{\mathbf{1}\text{ mile}} \equiv \frac{\mathbf{6080}\text{ ft}}{\mathbf{1}\text{ British nautical mile}} \equiv \frac{\mathbf{1852}\text{ m}}{\mathbf{1}\text{ international nautical mile}}$$

Area

$$1 \equiv \frac{\mathbf{144}\text{ in}^2}{\mathbf{1}\text{ ft}^2} \equiv \frac{0\cdot0929\text{ m}^2}{1\text{ ft}^2} \equiv \frac{645\text{ mm}^2}{1\text{ inch}^2} \equiv \frac{0\cdot836\text{ m}^2}{1\text{ yd}^2} \equiv \frac{\mathbf{4840}\text{ yd}^2}{\mathbf{1}\text{ acre}} \equiv \frac{\mathbf{640}\text{ acre}}{\mathbf{1}\text{ mile}^2}$$

$$\equiv \frac{0\cdot4047\text{ hectare}}{1\text{ acre}} \equiv \frac{4\cdot047 \times 10^3\text{ m}^2}{1\text{ acre}}$$

Velocity

$$1 \equiv \frac{0\cdot3048\text{ m/s}}{1\text{ ft/s}} \equiv \frac{0\cdot00508\text{ m/s}}{1\text{ ft/min}} \equiv \frac{1\cdot467\text{ ft/s}}{1\text{ mile/h}} \equiv \frac{\mathbf{88}\text{ ft/min}}{\mathbf{1}\text{ mile/h}} \equiv \frac{1\cdot609\text{ km/h}}{1\text{ mile/h}}$$

$$\equiv \frac{0\cdot515\text{ m/s}}{1\text{ (British) knot}} \equiv \frac{0\cdot447\text{ m/s}}{1\text{ mile/h}}$$

Second moment of area

$$1 \equiv \frac{8{\cdot}63 \times 10^{-3} \text{ m}^4}{1 \text{ ft}^4} \equiv \frac{4{\cdot}162 \times 10^{-7} \text{ m}^4}{1 \text{ in}^4}$$

Angle

$$1 \equiv \frac{\pi \text{ radians}}{180 \text{ degrees}} \equiv \frac{60 \text{ minutes of arc}}{1 \text{ degree}} \equiv \frac{60 \text{ seconds of arc}}{1 \text{ minute of arc}}$$

Volume; Modulus of section

$$1 \equiv \frac{1728 \text{ in}^3}{1 \text{ ft}^3} \equiv \frac{2{\cdot}832 \times 10^{-2} \text{ m}^3}{1 \text{ ft}^3} \equiv \frac{28{\cdot}32 \text{ litre}}{1 \text{ ft}^3} \equiv \frac{27 \text{ ft}^3}{1 \text{ yd}^3} \equiv \frac{1{\cdot}639 \times 10^{-5} \text{ m}^3}{1 \text{ in}^3}$$

$$\equiv \frac{0{\cdot}7646 \text{ m}^3}{1 \text{ yd}^3} \equiv \frac{0{\cdot}1605 \text{ ft}^3}{1 \text{ Imperial gallon}} \equiv \frac{4{\cdot}546 \times 10^{-3} \text{ m}^3}{1 \text{ Imperial gallon}} \equiv \frac{231 \text{ in}^3}{1 \text{ U.S. gallon}}$$

$$\equiv \frac{0{\cdot}1337 \text{ ft}^3}{1 \text{ U.S. gallon}} \equiv \frac{0{\cdot}833 \text{ Imperial gallon}}{1 \text{ U.S. gallon}}$$

Mass

$$1 \equiv \frac{16 \text{ ozm}}{1 \text{ lbm}} \equiv \frac{112 \text{ lbm}}{1 \text{ cwt}} \equiv \frac{2240 \text{ lbm}}{1 \text{ tonm}} \equiv \frac{2000 \text{ lbm}}{1 \text{ U.S. tonm}} \equiv \frac{0{\cdot}4536 \text{ kg}}{1 \text{ lbm}} \equiv \frac{1016 \text{ kg}}{1 \text{ tonm}}$$

$$\equiv \frac{32{\cdot}17 \text{ lbm}}{1 \text{ slug}} \equiv \frac{14{\cdot}59 \text{ kg}}{1 \text{ slug}}$$

Time interval

$$1 \equiv \frac{60 \text{ s}}{1 \text{ min}} \equiv \frac{60 \text{ min}}{1 \text{ h}} \equiv \frac{3600 \text{ s}}{1 \text{ h}} \equiv \frac{24 \text{ h}}{1 \text{ day}} \equiv \frac{1440 \text{ min}}{1 \text{ day}} \equiv \frac{86400 \text{ s}}{1 \text{ day}}$$

$$\equiv \frac{365{\cdot}24 \text{ solar days}}{1 \text{ year}}$$

Volume flow rate

$$1 \equiv \frac{28{\cdot}32 \times 10^{-3} \text{ m}^3/\text{s}}{1 \text{ ft}^3/\text{s}} \equiv \frac{101{\cdot}9 \text{ m}^3/\text{h}}{1 \text{ ft}^3/\text{s}} \equiv \frac{373{\cdot}7 \text{ Imperial gallons}/\text{min}}{1 \text{ ft}^3/\text{s}}$$

$$\equiv \frac{2{\cdot}242 \times 10^4 \text{ Imperial gallons}/\text{h}}{1 \text{ ft}^3/\text{s}} \equiv \frac{7{\cdot}577 \times 10^{-5} \text{ m}^3/\text{s}}{1 \text{ Imperial gallon}/\text{min}}$$

$$\equiv \frac{2{\cdot}676 \times 10^{-3} \text{ ft}^3/\text{s}}{1 \text{ Imperial gallon}/\text{min}}$$

Density

$$1 \equiv \frac{16{\cdot}02 \text{ kg}/\text{m}^3}{1 \text{ lbm}/\text{ft}^3} \equiv \frac{62{\cdot}4 \text{ lbm}/\text{ft}^3}{1000 \text{ kg}/\text{m}^3} \equiv \frac{1000 \text{ kg}/\text{m}^3}{1 \text{ g}/\text{cm}^3}$$

Moment of inertia (Second moment of mass)

$$1 \equiv \frac{0{\cdot}04214 \text{ kg m}^2}{1 \text{ lbm ft}^2} \equiv \frac{2{\cdot}926 \times 10^{-4} \text{ kg m}^2}{1 \text{ lbm in}^2} \equiv \frac{1{\cdot}356 \text{ kg m}^2}{1 \text{ slug ft}^2}$$

Force

$$1 \equiv \frac{32\cdot17 \text{ pdl}}{1 \text{ lbf}} \equiv \frac{10^{-5} \text{ N}}{1 \text{ dyne}} \equiv \frac{0\cdot1383 \text{ N}}{1 \text{ pdl}} \equiv \frac{4\cdot448 \text{ N}}{1 \text{ lbf}} \equiv \frac{0\cdot4536 \text{ kgf}\dagger}{1 \text{ lbf}} \equiv \frac{9\cdot81 \text{ N}}{1 \text{ kgf}\dagger}$$
$$\equiv \frac{9964 \text{ N}}{1 \text{ tonf}}$$

Moment of force, or torque

$$1 \equiv \frac{1\cdot356 \text{ N m}}{1 \text{ lbf ft}} \equiv \frac{3037 \text{ N m}}{1 \text{ tonf ft}}$$

Pressure; Stress

$$1 \equiv \frac{144 \text{ lbf/ft}^2}{1 \text{ lbf/in}^2} \equiv \frac{6895 \text{ N/m}^2}{1 \text{ lbf/in}^2} \equiv \frac{0\cdot0703 \text{ kgf/cm}^2}{1 \text{ lbf/in}^2} \equiv \frac{1\cdot544 \times 10^7 \text{ N/m}^2}{1 \text{ tonf/in}^2}$$
$$\equiv \frac{1\cdot013 \times 10^5 \text{ N/m}^2}{1 \text{ atm}} \equiv \frac{14\cdot70 \text{ lbf/in}^2}{1 \text{ atm}} \equiv \frac{2116 \text{ lbf/ft}^2}{1 \text{ atm}} \equiv \frac{10^5 \text{ N/m}^2}{1 \text{ bar}}$$
$$\equiv \frac{2\cdot036 \text{ in Hg}\ddagger}{1 \text{ lbf/in}^2} \equiv \frac{29\cdot92 \text{ in Hg}\ddagger}{1 \text{ atm}} \equiv \frac{7\cdot50 \times 10^{-3} \text{ mm Hg}\ddagger}{1 \text{ N/m}^2} \equiv \frac{0\cdot491 \text{ lbf/in}^2}{1 \text{ in Hg}\ddagger}$$
$$\equiv \frac{3386 \text{ N/m}^2}{1 \text{ in Hg}\ddagger} \equiv \frac{133\cdot3 \text{ N/m}^2}{1 \text{ mm Hg}\ddagger} \equiv \frac{2\cdot307 \text{ ft water}\S}{1 \text{ lbf/in}^2} \equiv \frac{249\cdot1 \text{ N/m}^2}{1 \text{ in water}\S}$$
$$\equiv \frac{0\cdot03613 \text{ lbf/in}^2}{1 \text{ in water}\S} \equiv \frac{2989 \text{ N/m}^2}{1 \text{ ft water}\S}$$

Work; Energy

$$1 \equiv \frac{1\cdot356 \text{ J}}{1 \text{ ft lbf}} \equiv \frac{3\cdot766 \times 10^{-7} \text{ kW h}}{1 \text{ ft lbf}} \equiv \frac{3\cdot6 \times 10^6 \text{ J}}{1 \text{ kW h}} \equiv \frac{1055 \text{ J}}{1 \text{ Btu}\|}$$
$$\equiv \frac{2\cdot931 \times 10^{-4} \text{ kW h}}{1 \text{ Btu}\|} \equiv \frac{252\cdot0 \text{ cal}\|}{1 \text{ Btu}\|} \equiv \frac{1\cdot8 \text{ Btu}}{1 \text{ CHU}} \equiv \frac{10^5 \text{ Btu}}{1 \text{ therm}} \equiv \frac{778 \text{ ft lbf}}{1 \text{ Btu}\|}$$
$$\equiv \frac{1401 \text{ ft lbf}}{1 \text{ CHU}} \equiv \frac{4\cdot187 \text{ J}}{1 \text{ cal}\|} \equiv \frac{1\cdot163 \times 10^{-6} \text{ kW h}}{1 \text{ cal}\|} \equiv \frac{1\cdot602 \times 10^{-19} \text{ J}}{1 \text{ eV}}$$

Power

$$1 \equiv \frac{1\cdot356 \text{ W}}{1 \text{ ft lbf/s}} \equiv \frac{550 \text{ ft lbf/s}}{1 \text{ hp}} \equiv \frac{75 \text{ m kgf/s}}{1 \text{ C.V. (metric horsepower)}}$$
$$\equiv \frac{0\cdot986 \text{ hp}}{1 \text{ C.V. (metric horsepower)}} \equiv \frac{3\cdot412 \text{ Btu/h}}{1 \text{ W}} \equiv \frac{746 \text{ W}}{1 \text{ hp}}$$

(Absolute or dynamic) Viscosity

$$1 \equiv \frac{47\cdot9 \text{ N s/m}^2}{1 \text{ lbf s/ft}^2} \equiv \frac{0\cdot1 \text{ N s/m}^2}{1 \text{ poise}} \equiv \frac{1\cdot488 \text{ N s/m}^2}{1 \text{ pdl s/ft}^2}$$

† Known in Germany and Eastern Europe as kilopond (kp).
‡ Strictly for mercury at 0°C and 1 atm pressure.
§ Strictly for pure water at 4°C and 1 atm pressure.
‖ The International Table value.

Kinematic viscosity (μ/ρ)

$$1 \equiv \frac{0\cdot0929 \text{ m}^2/\text{s}}{1 \text{ ft}^2/\text{s}} \equiv \frac{10^{-4} \text{ m}^2/\text{s}}{1 \text{ stokes}}$$

Electric charge

$$1 \equiv \frac{10 \text{ coulomb}}{1 \text{ abcoulomb}} \equiv \frac{1 \text{ coulomb}}{2\cdot998 \times 10^9 \text{ statcoulomb}}$$

Electric current

$$1 \equiv \frac{10 \text{ ampere}}{1 \text{ abampere}} \equiv \frac{1 \text{ ampere}}{2\cdot998 \times 10^9 \text{ statampere}}$$

Electric potential

$$1 \equiv \frac{10^{-8} \text{ volt}}{1 \text{ abvolt}} \equiv \frac{299\cdot8 \text{ volt}}{1 \text{ statvolt}}$$

Electric resistance

$$1 \equiv \frac{10^{-9} \text{ ohm}}{1 \text{ abohm}} \equiv \frac{8\cdot987 \times 10^{11} \text{ ohm}}{1 \text{ statohm}}$$

Electric capacitance

$$1 \equiv \frac{10^9 \text{ farad}}{1 \text{ abfarad}} \equiv \frac{1 \text{ farad}}{8\cdot987 \times 10^{11} \text{ statfarad}}$$

Magnetic flux

$$1 \equiv \frac{10^{-8} \text{ weber}}{1 \text{ maxwell}}$$

Magnetomotive force

$$1 \equiv \frac{10 \text{ ampere-turn}}{4\pi \text{ gilbert}}$$

Inductance

$$1 \equiv \frac{10^{-9} \text{ henry}}{1 \text{ abhenry}} \equiv \frac{8\cdot987 \times 10^{11} \text{ henry}}{1 \text{ stathenry}}$$

Magnetic flux density

$$1 \equiv \frac{10^{-4} \text{ tesla}}{1 \text{ gauss}} \equiv \frac{10^{-9} \text{ tesla}}{1 \text{ gamma}}$$

Magnetic field strength

$$1 \equiv \frac{10^3 \text{ A/m}}{4\pi \text{ oersted}}$$

Illumination

$$1 \equiv \frac{10\cdot76 \text{ lux}}{1 \text{ fc}}$$

Luminance

$$1 \equiv \frac{10^4 \text{ cd/m}^2}{\pi \text{ lambert}} \equiv \frac{3\cdot426 \text{ cd/m}^2}{1 \text{ ft-lambert}} \equiv \frac{\pi \text{ apostilb}}{1 \text{ cd/m}^2}$$

6

The Form of Expressions in Physical Algebra

6.1 Types of Magnitudes and Equations

As we saw in Section 1.2, any formula relating the magnitudes of physical quantities is more than a piece of 'ordinary' algebra. The magnitude of a physical quantity is expressed by the product of a numeric and a unit, the unit being a suitable sample of that kind of quantity. Therefore any relation connecting the magnitudes of physical quantities involves not only the numerics but also the units.

A physical quantity is something that can be measured by some strictly definable process. In general, a given physical quantity may be measured in a number of ways; length, for example, is most commonly measured by the use of a calibrated rule, but it may also be measured by triangulation or by timing the movement of a body having a constant velocity or by counting optical interference fringes. If results of such different measurement processes are not exactly compatible, then final appeal is by universal convention made to one particular method (in the case of length this is measurement by calibrated rule).

In principle, the measurement of a physical quantity consists of comparison with a standard amount of that quantity termed the unit. The result of the measurement is known as the magnitude of the quantity. For many quantities, however, direct comparison with the unit is made only with extreme difficulty, if at all. Consequently, the magnitude of such a quantity is determined by the measurement of other quantities and the required magnitude is then calculated from appropriate formulae. For example, the surface tension of a liquid is frequently 'measured' by the direct measurements of the diameter of a capillary tube and the rise of the liquid up the tube; the magnitude of the surface tension is then calculated from a formula. (If the diameter of the tube is extremely small, the formula is $\gamma = \rho g h d/(4 \cos \theta)$. Here γ represents the magnitude of the surface tension, ρ the magnitude of the density of the liquid, g the weight per unit mass, h the magnitude of the 'capillary rise', d the magnitude of the diameter of the tube, and θ the magnitude of the angle of contact between the liquid surface and the wall of the tube.) The term 'measurement' then includes not only the operations of comparison with a unit (i.e. direct measurement), but also any subsequent mathematical operations required.

Magnitudes of physical quantities may be regarded as either *fundamental* or *derived*. Fundamental magnitudes (also termed *primary* magnitudes) are those which are not defined in terms of the magnitude of any other physical quantity. For example, if length is measured by comparison with a calibrated rule, then the magnitude of any length is a fundamental magnitude because it is determined solely by the number of units of length found in the corresponding length on the rule. Methods of measurement and sizes of units used for other quantities such as time interval, mass, acceleration and so on could be altered, but such alterations would have no effect on the expression for the magnitude of length.

A derived (or *secondary*) magnitude, on the other hand, *is* defined in terms of the magnitudes of other quantities, and, in general, the expression of a derived magnitude is altered by changes in the units of other quantities. For example, velocity may be measured in various ways, but the basic, canonical, method of determining its magnitude is by dividing the magnitude of a length by the magnitude of a time interval. Thus if the size of either the unit of length or the unit of time interval is altered, the expression for the magnitude of a velocity is changed. The magnitude of velocity, then, according to the usual conventions is a derived magnitude since it is defined in terms of the magnitudes of other quantities.

Similarly, density is a quantity with a derived magnitude. Except in a few special instances, one cannot directly compare the density of one substance with a standard density chosen as the unit of density. Instead, one determines the mass of a certain amount of the substance in question and also its volume: the magnitude of the density is then determined by calculation. Suitable units for density are kg/m^3 or lbm/ft^3; thus in this case the fundamental units (those of mass and length) used in defining the derived unit are explicitly mentioned. However, they need not be. There is no reason why a separate name should not be chosen for a unit of density—the 'squash' perhaps. Yet so long as the squash has to be defined in terms of units of mass and volume (or length) it remains a derived magnitude in spite of its distinctive name.

The classification of magnitudes as fundamental and derived is not a rigid one imposed by restrictions of nature, but is to a large extent arbitrary. For example, the magnitude of a force is normally defined in terms of chosen units of mass and acceleration. It is, however, quite possible to compare one force directly with another, and so the magnitude of a force could be regarded as fundamental. It would then be possible, by reversing the usual defining procedure, to regard the magnitude of mass as derived. (Mass = Force ÷ Acceleration.) There are other quantities too, whose magnitudes are normally regarded as derived, which could be considered to have fundamental magnitudes. The number of fundamental magnitudes is in fact arbitrary: there is no reason why, with sufficient ingenuity, any magnitude should not be definable fundamentally, that is, without reference to any other magnitude.

The magnitude of any physical quantity may be related to the magnitudes of other physical quantities primarily by proportionalities deduced from experimental observations. It is a matter of experimental observation, for example, that the magnitude of the net force acting on a body of fixed mass is proportional to the product of the magnitudes of the mass of that body and the acceleration produced. These proportional relationships may be put into the form of equations by introducing constants of proportionality. The values of these constants, however, remain undetermined until the magnitudes of the quantities are in some way defined. If the magnitudes are fundamental ones they are defined quite separately from the equations: the magnitude of mass, for example, may be specified by comparison with the block of matter known as the International Prototype Kilogram. The definition of a derived magnitude, however, requires the use of an equation. Any equation in which that magnitude appears may serve to define it if a particular value (for simplicity usually unity) is chosen for the constant of proportionality.

The magnitude of force, for example, is by convention defined by putting $K = 1$ in the expression $F = Kma$. The resulting equation $F = ma$ may thus be termed the *defining equation* for the derived magnitude of force.

We may note that any other relation involving the magnitude of force could have been used to define it. Once the magnitude has been defined by $F = ma$, however, the constants of proportionality are no longer arbitrary in other expressions relating the magnitude of force to fundamental ones. Newton's Law of Universal Gravitation, for example, states that the force F of gravitational attraction between two bodies of mass m_1 and m_2 separated by a distance r (large compared with the bodies) is proportional to $m_1 m_2/r^2$. Having already defined the magnitude of F, however, we are obliged to write the law in the form $F = Gm_1m_2/r^2$. Here the constant G has a definite value which may be calculated from a set of corresponding values of F, m_1, m_2 and r. Then $G = Fr^2/m_1m_2$. If the Law of Universal Gravitation had been chosen as the defining equation for force, G would (probably) have been unity, but a constant K (not equal to unity) would then have been necessary in the inertial equation $F = Kma$.

Equations relating the magnitudes of physical quantities are thus of two kinds: (a) those which define derived magnitudes ('defining' equations), and (b) those which express physical relations which have to be established by measurement. The equation $F = ma$ is of type (a) because it defines the magnitude of force in terms of those of mass and acceleration. The equation $F = Gm_1m_2/r^2$, however, is of type (b) because measurements of F, m_1, m_2 and r are required to establish the value of G. Whether a particular equation is of one kind or the other depends on which magnitudes have been chosen as fundamental, and on how the derived magnitudes are defined from the fundamental ones.

6.2 The Nature of Dimensional Formulae

Magnitudes of physical quantities, as we have seen, are expressed by the product of a numeric and a unit, and it is the unit which essentially distinguishes the magnitude of one kind of quantity from that of another. We now introduce the symbol '[X]', which, for the moment, will be defined as 'a unit of X', X being some physical quantity. Thus the symbol [L] may be used to represent a unit of length (for brevity we use the initial letter L for the quantity length). Similarly [T] may represent a unit interval of time, [M] a unit mass and so on.

If the magnitudes of length and time interval are regarded as fundamental, the magnitude of velocity may be regarded as derived. The definition of the magnitude of velocity, v, then comes from the defining equation $v = l/t$, where l represents the magnitude of the length traversed by a moving object and t the magnitude of the corresponding time interval (assumed small enough for the

§6.2]   THE NATURE OF DIMENSIONAL FORMULAE   51

velocity to be considered uniform during that interval). In a particular instance, the magnitude of the length would be expressed by $n_1[L]$, and that of the time interval by $n_2[T]$, where n_1 and n_2 are numerics. Thus the magnitude v is given by $n_1[L]/n_2[T]$. As n_1/n_2 is a ratio of numbers it may give place to another numeric, say n_3. Consequently

$$v = n_3 \frac{[L]}{[T]} \qquad (6.1)$$

Since a velocity, like other physical quantities, has its magnitude expressed by the product of a numeric and a unit, eqn 6.1 shows that the expression $[L]/[T]$ represents a possible unit of velocity. That is, a unit of velocity has been derived from units of length and time interval. One could alternatively use an entirely arbitrary unit of velocity, but the magnitude of velocity would then be fundamental, not derived, and the experimental fact that velocity is proportional to the length traversed and inversely proportional to the corresponding time interval would have to be expressed by the equation $v = Kl/t$.

The expression $[L]/[T]$ is usually known as the *dimensional formula* of velocity —although it should be noted that the dimensional formula is characteristic not of velocity itself but of the units with which its magnitude is expressed. A further step gives the dimensional formula of acceleration. Accepting the equation $a = dv/dt$ as the defining equation for the magnitude of acceleration, we may suppose that the increase of velocity is expressed by $n_1[V]$ and that this occurs during a (short) time interval of magnitude $n_2[T]$. The magnitude of the acceleration is then $n_1[V]/n_2[T] = n_3[V]/[T]$, say. Thus $[V]/[T]$ represents a possible unit of acceleration. However, $[V]$, a unit of velocity, has itself been considered a derived magnitude. With the substitution $[V] \equiv [L]/[T]$ the dimensional formula of acceleration becomes $([L]/[T]) \div [T]$ which may be contracted to $[L]/[T]^2$ or $[L][T]^{-2}$.

In practice, units of different size are often used to express the magnitude of the same quantity. The symbol $[X]$ therefore refers only to a *possible* unit of X; it is not restricted to any particular unit. Indeed, in a particular calculation of acceleration, for example, two units of time interval could be used: the increase of velocity could be expressed in metres per second, but the time interval over which it occurs might be expressed in minutes. Such a result as 120 m/s per minute is, however, identical with

$$120 \frac{\text{m}}{\text{s min}} \times \frac{1 \text{ min}}{60 \text{ s}} = 2 \text{ m/s}^2$$

and so the dimensional formula $[L][T]^{-2}$ is still appropriate.

Since the magnitude of force is defined by the equation $F = ma$, the dimensional formula of force is $[M] \times [L][T]^{-2}$. By similar reasoning the dimensional formulae of other magnitudes may be built up from those of fundamental magnitudes. The power to which a fundamental unit is raised in the dimensional formula of any magnitude is said to be the dimension in respect to that fundamental magnitude in the formula. For example, the magnitude of force is said to have dimensions of 1 in respect to mass, 1 in respect to length and -2 in respect to time interval. (Although this usage is generally understood it is not always followed. What is here termed the dimensional formula is often loosely called 'the dimensions'.)

Dimensional formulae are characteristic of the *magnitudes* of physical quantities but not of the quantities themselves. It is often wrongly supposed that dimensional formulae in some way give information about the intrinsic nature of the quantities with which they are associated. However, a dimensional formula for the magnitude of a particular quantity is not unique: it depends on which magnitudes are regarded as fundamental and on how the derived magnitudes are defined. For this reason, care has been taken here to refer to the dimensional formula of the magnitude or of the unit. To say that the dimensional formula associated with viscosity, for example, is $[M][L]^{-1}[T]^{-1}$ tells us nothing at all about the intrinsic nature of viscosity—the 'lack of slipperiness', as Newton put it. This kind of information must come from the *definition* of the quantity, not from a dimensional formula. The dimensional formula tells us only how the magnitude of the quantity concerned is defined in terms of fundamental magnitudes; in other words, it defines a possible unit of that quantity. From this it follows that any magnitude which may be expressed by only a numeric has no dimensions, that is, it is 'dimensionless' or 'non-dimensional'.

Apart from the exponents (the 'dimensions'), numbers find no place in dimensional formulae. Although the kinetic energy of a particle of mass m moving with velocity v is given by $\frac{1}{2}mv^2$, the dimensional formula excludes the $\frac{1}{2}$ since that is simply included with the numerics of m and v in the arithmetical calculations.

For brevity, only one pair of square brackets is usually employed in writing a dimensional formula. That for viscosity, for example, is written $[ML^{-1}T^{-1}]$, and the brackets may now be taken to mean 'dimensional formula herein'. Other abbreviations are also made. The phrase 'the magnitude of' is frequently omitted. Thus we say 'μ is the viscosity' when we really mean 'the symbol μ represents the magnitude of the viscosity'.

Although the magnitudes of length, mass and time interval are usually regarded as fundamental, they are in no way superior to other magnitudes. It is true that in selecting units for actual measurements practical considerations arise (as we saw in Section 3.1) which have led to magnitudes of length, mass and time interval being used as standards. In considering dimensional formulae, however, our concern is not primarily with *actual* measurements. There is thus no need to regard magnitudes of length, mass and time interval, or any other particular magnitudes, as fundamental for this purpose. Nevertheless, for the majority of applications it is most convenient to consider the magnitudes of length, mass and time interval as fundamental, and other magnitudes in mechanics as derived from them.

Other fundamental magnitudes are usually required when quantities connected with heat, electricity, magnetism or light are under discussion. Examination of these will be left for a while so that we may establish the principles of dimensional analysis without too much complication.

6.3 Dimensional Homogeneity

It may be recalled from Section 1.2 that if magnitudes of physical quantities are to be added, subtracted, or equated, the quantities must be of the same kind, because only then do these processes have any physical significance. Now, for a given choice of fundamental magnitudes, quantities of the same kind have magnitudes with the same dimensional formulae, and any equation involving the magnitudes must be *dimensionally homogeneous*.

On this principle of dimensional homogeneity rests the whole of the mathematics associated with physics and engineering. It cannot be avoided if relations connecting the magnitudes of physical quantities are to be expressed in mathematical form.

The principle is simply that the dimensional formulae of magnitudes which are added, subtracted or equated must be identical. A word of warning is perhaps necessary here. Once a particular set of fundamental magnitudes ([L], [M] and [T], for example) has been chosen, then if two quantities are of the same kind it is a necessary consequence that the dimensional formulae of their magnitudes are identical. The converse statement, however,—that the same dimensional formulae imply the same kind of quantity—is not necessarily true. When, as is usual for the sake of simplicity, a small number of magnitudes are considered fundamental, it occasionally happens by coincidence that two different derived magnitudes have identical dimensional formulae. A common instance concerns the quantities torque and work or energy. The magnitude of each is normally defined as the product of magnitudes of force and distance; thus the dimensional formula of each is [F][L] or [MLT^{-2}][L] $\equiv$ [ML^{2}T^{-2}]. Yet, whereas for defining torque the force and the distance are perpendicular, for work (or energy) the force and the distance are in the same direction. Consequently, torque and energy are not at all similar. To attempt to add the magnitudes of a torque and an amount of energy, for example, would be just as illogical and futile as attempting to add the magnitudes of a mass and a length.

Applied to a defining equation the principle of dimensional homogeneity is a truism because the dimensional formulae of the two sides are made identical by the definition. Any other equation of 'physical algebra' must, however, also be dimensionally homogeneous if it is to be correct and to have physical significance.

In addition to the variables on which interest is centred, physical equations may contain constants. In a defining equation such a constant is usually simply a number. It will have been chosen arbitrarily; in many cases, although not in all, it will have been chosen as unity. At all events, being a number it is itself dimensionless. In other equations, however, the constants which have to be included if the equations are to balance must be determined ultimately by experiment. They are not in general dimensionless; their dimensional formulae are determined from those of the other magnitudes in the equation so that dimensional homogeneity is achieved.

For example, in Newton's Law of Universal Gravitation, $F = Gm_1m_2/r^2$, the constant G must have the same dimensional formula as Fr^2/m_1m_2, that is [MLT^{-2}][L^2]/[M][M] $\equiv$ [L^{3}M^{-1}T^{-2}], otherwise the equation would not be dimensionally homogeneous. The fact that G is a 'universal constant' is irrelevant: dimensions are associated with it, and in analysing the equation they must be accounted for.

An important consequence of the principle of dimensional homogeneity is that the truth of a correct physical relation in which all relevant factors are explicitly included does not depend on the size of units used in expressing the various magnitudes. If a relation is dimensionally homogeneous, the same fundamental units appear to the same power in the dimensional formulae of each side of the relation. Thus if the size of any of the fundamental units is changed by some factor, each side of the relation is similarly affected and the relation is no less true than formerly.†

The principle of dimensional homogeneity, as we have seen, is a condition for a relation in 'physical algebra' to have physical significance. Although a relation may be dimensionally homogeneous yet wrong in some other respect, a relation which is *not* dimensionally homogeneous cannot have physical meaning, and an error has probably been made in its derivation. Checking relations for dimensional homogeneity should become an automatic habit. If the dimensional formulae of terms are verified at frequent intervals in algebraic working, many slips may be detected and corrected.

The general concepts of dimensional formulae and dimensional homogeneity have been discussed in some detail here because a clear understanding of them is essential if the maximum advantage is to be taken of the methods of dimensional analysis. It is a subject to which philosophers of science have devoted much attention, if not always clarity of thought, and agreement has yet to be reached on finer points of the theory. Of the great value of the methods of dimensional analysis, however, there is no doubt, and here we have sought only to establish a firm basis from which these methods may be used.

6.4 Dimensional Analysis

The principle of dimensional homogeneity enables us to test the consistency and often the completeness of an equation in 'physical' algebra. More important, however, the requirement of dimensional homogeneity also imposes conditions on the quantities involved in a physical problem, and so provides valuable clues to the form of the relation connecting their magnitudes. This is because a relation can, in general, be dimensionally homogeneous only if the individual magnitudes enter it in certain definite combinations. Knowledge of the form of the relation is valuable not only in interpreting the results of experiments, but also in suggesting the pattern which a series of experiments could most usefully take. The search for the correct form of the relation is known as dimensional analysis.

† For some practical calculations formulae may be used which, although 'correct', are incomplete in the sense that not all relevant variables are explicitly included. An example of this sort of relation is the widely used Manning formula which describes the flow of a liquid in an open channel (such as a river or canal) when the free surface of the liquid is parallel to the base of the channel. If the cross-section of the flow has an area A m² and a perimeter P m in contact with the solid boundaries, and if the mean velocity of the liquid over this cross-section is v m/s, then

$$v = (A/P)^{2/3} i^{1/2}/n$$

where i represents the downward slope of the base (i.e. vertical drop divided by corresponding length along channel) and n is a number dependent on the roughness of the boundary surface. This formula explicitly includes neither the viscosity nor the density of the liquid, yet each of these properties can significantly affect the flow. Such 'incomplete' formulae are frequently limited in application to a certain range of the variables involved (usually that corresponding to the experimental data from which the formulae were derived). Moreover, they are 'ordinary', not 'physical' algebra and so may be true only if a particular set of units is used (metres and seconds in the example just quoted).

§6.4] DIMENSIONAL ANALYSIS

The methods of dimensional analysis are of very wide application and are especially valuable in the study of phenomena which are too complex for complete theoretical treatment. Such complexity may result from a large number of independent variables affecting the phenomenon, or from a mathematically complicated form of the expressions relating these variables. Dimensional analysis alone can never give the complete solution of a problem. It does, however, usually permit considerable simplifications in investigating complex phenomena and may show the effect of particular variables, especially when the effects of some of the other variables are known.

6.4.1 The Process of Analysis

In the most general terms the problem to be solved may be stated thus: We are interested in a certain quantity (for example, a particular force or the velocity of a particular particle in a specified system, or the power required for a particular purpose). How does the magnitude of this quantity depend on the magnitude of other quantities which could affect the situation?

In other words, the problem is to obtain a relation of the form

$$Q_1 = \phi(Q_2, Q_3, Q_4, \ldots) \tag{6.2}$$

Here Q_1 represents the magnitude of the quantity of especial interest; Q_2, Q_3, Q_4, etc. the magnitudes of the other quantities entering the problem. The symbol ϕ means 'some function of'. Nothing is implied about the form which the function may take: it involves the magnitudes $Q_2, Q_3, Q_4, \ldots$ in some way but not even necessarily in the form of a product.

Now it is a fundamental proposition of mathematics[†] that any sort of function whatever, provided only that it is continuous, may be represented to as close an approximation as desired by a series of terms, each of which is the product of a numeric and powers of the arguments of the function. Thus the function $\phi(Q_2, Q_3, Q_4)$, for example, may be expressed as the series

$$k_1 Q_2^{a_1} Q_3^{b_1} Q_4^{c_1} + k_2 Q_2^{a_2} Q_3^{b_2} Q_4^{c_2} + k_3 Q_2^{a_3} Q_3^{b_3} Q_4^{c_3} + \ldots$$

where $k_1, k_2, k_3, \ldots$ are numerics (and so dimensionless). By the principle of dimensional homogeneity all terms in this series must have the same dimensional formula. Moreover, in eqn 6.2, the dimensional formula of Q_1 must be the same as the dimensional formula of each term of the series expressing the right-hand side. Thus, using square brackets to mean 'the dimensional formula of', we have

$$[Q_1] \equiv [Q_2^a Q_3^b Q_4^c \ldots] \tag{6.3}$$

(The numerics k, being dimensionless, do not enter dimensional formulae.)

Our first task in carrying out the analysis is clearly to decide what are the quantities whose magnitudes Q_2, Q_3, Q_4, etc. enter the problem. It is at this point particularly that there is risk of error. No quantity which might affect the result must be overlooked even though, in the conditions of the problem considered, that quantity might not vary in magnitude. For instance, if one of the factors entering the problem is the gravitational force tending to pull a particular body towards the earth, then the weight per unit mass, g, must be included in the list of significant quantities—even though its magnitude is constant at, say, 9·81 N/kg, and so it is not, in the ordinary sense, a 'variable'. In some problems

[†] Weierstrass' Approximation Theorem.

dimensional constants appear—such as the gravitational constant G or the velocity of light. If they are relevant to the problem under investigation they must be included in the analysis because their magnitudes are not merely numbers. The fact that their magnitudes do not vary is beside the point.

On the other hand, there is no virtue in bringing in quantities which have no bearing on the situation. For example, in a problem concerned with the equilibrium of a fluid, the viscosity of the fluid—which is in evidence only when there is relative motion between different particles of fluid—need not be considered. Moreover, some quantities have only an indirect effect on a problem. Temperature, for example, may affect the magnitude of some of the quantities which directly enter a physical relation—such as density, electrical resistance, viscosity and other properties of substances. But the effect of a variation in the magnitude of the property could just as well be achieved by changing the substance instead of the temperature. From the point of view of dimensional analysis, therefore, temperature may in such circumstances be left out of account because its effect is confined to determining values of variables which directly influence the phenomenon being investigated; consequently, the way in which the magnitude of the temperature might be expressed (whether on a scale of degrees Celsius, for example, or on an absolute or on any other scale) is here irrelevant. In problems of heat transmission, however, the temperature of a body can have a direct effect on the phenomenon and so can itself be expected to appear in the sought-for relation. In this instance, then, temperature must be included in the list of relevant quantities.

6.4.2 Rayleigh's Method

6.4.2.1 The Simple Pendulum

To demonstrate the basic method of dimensional analysis we first consider a simple problem in mechanics. It is one to which the solution is known by other means and so the soundness of the method may be readily verified. The investigation concerns the oscillations of a simple pendulum and we suppose that the quantity of primary interest is the time of swing, t. What are the quantities which may affect the time of swing?

The length of the pendulum, l, and the mass of the bob, m, can both be expected to have an effect. The downward movement of the bob to its lowest point is a result of the force pulling it towards the earth: therefore the weight per unit mass, g, is involved. The amplitude of the swing may also have something to do with the matter; this we may represent by the angle α through which the pendulum oscillates on each side of the vertical. If the effect of friction (both mechanical and of the atmosphere) is negligible, there is no other quantity which can be expected to influence the time of swing. The relation sought is consequently of the form

$$t = \phi(l, m, g, \alpha) \tag{6.4}$$

In this arrangement of terms, t is isolated on one side of the equation, and thus is called the *dependent* variable (since its value depends on the values of the other terms). The terms l, m, g and α are described as the *independent* variables because any one of them could be changed in magnitude without affecting the others; for example, if l were altered in magnitude, only t would be affected and not the

§6.4]   DIMENSIONAL ANALYSIS

other independent variables m, g, α. However, the words dependent and independent refer only to the mathematical arrangement, and do not carry any implication of physical cause and effect. After all, by algebraic rearrangement any variable could be isolated on one side of an equation and thus made the dependent variable. So, although in eqn 6.4 t has been made the dependent variable, it has no special status.

Now the dimensional formula of t must be the same as that of the as yet unknown function on the right-hand side of eqn 6.4. So, by the arguments which led to eqn 6.3, in which we consider one typical term of the series expressing the function, we have

$$[t] \equiv [l^a m^b g^c \alpha^d]$$

If, as is usual, the magnitudes of length, mass and time interval are regarded as fundamental, the variables t, l and m in the present problem are fundamental magnitudes and their dimensional formulae are simply [T], [L] and [M], respectively. The dimensional formula of g is that corresponding to weight (i.e. force) divided by mass, that is to acceleration, $[LT^{-2}]$.

When only the magnitudes of length, mass and time interval are regarded as fundamental, the magnitude of angle must be defined as length of arc divided by radius.† The corresponding dimensional formula is therefore $[L]/[L] \equiv [L]^0 \equiv [1]$; in other words, the magnitude of angle is dimensionless. The relation between the dimensional formulae then becomes

$$[T] \equiv [L]^a [M]^b [LT^{-2}]^c [1]^d$$

Since the fundamental magnitude of mass appears not at all on the left-hand side of the relation and only once on the right-hand side, dimensional homogeneity in respect to mass can be achieved only if the exponent b is zero. Thus the mass of the pendulum bob has no effect on the time of swing—a fact amply verified by experiment.

As for dimensions in respect to time interval, [T] appears on the right-hand side to the power $-2c$ and on the left-hand side to the power 1. Therefore $-2c$ must equal 1, i.e. $c = -\frac{1}{2}$. Finally, [L] does not appear at all on the left-hand side but to the power $a + c$ on the right. For dimensional homogeneity, therefore, $a + c$ must be zero. Since c has already been determined as $-\frac{1}{2}$, a must be $+\frac{1}{2}$.

The analysis thus gives definite values of a, b and c. These have been found by equating dimensions in respect to the fundamental magnitudes of length, mass and time interval on the two sides of the equation. Such methods yield no information about the exponent d. But at least it may now be said that t may be expressed as a series of terms of the form $k_n l^a m^b g^c \alpha^d$, i.e. $k_n l^{1/2} m^0 g^{-1/2} \alpha^d$, i.e.

$$\left(\frac{l}{g}\right)^{1/2} (k_n \alpha^d)$$

The $(l/g)^{1/2}$ is clearly a factor of the series and so

$$t = \left(\frac{l}{g}\right)^{1/2} (k_1 \alpha^{d_1} + k_2 \alpha^{d_2} + k_3 \alpha^{d_3} + \ldots)$$

Using the fundamental mathematical axiom 'in reverse' we may write the series of α terms as $\phi_1(\alpha)$. The symbol ϕ_1 means 'some function of' and the suffix 1 has

† Other possibilities will be discussed in Section 6.5.

been used here to emphasize that the form of the function is not the same as that in eqn 6.4. The resulting formula is therefore

$$t = \left(\frac{l}{g}\right)^{1/2} \phi_1(\alpha) \tag{6.5}$$

This is as far as simple dimensional analysis will take us. No definite information about the form of the function $\phi_1(\alpha)$ can be obtained by considering dimensions. This is hardly surprising when it is realized that the argument of the function has been regarded as dimensionless. Three important things, however, have been discovered: (1) that the time of swing is independent of the mass of the pendulum; (2) that the time is proportional to the square root of the length; (3) that the time is inversely proportional to the square root of g.

The problem just considered is one for which the usual methods of mathematics yield a solution and so dimensional analysis is not there essential. The algebraic solution agrees with eqn 6.5 and shows further that, if the amplitude α is small, $\phi_1(\alpha)$ is a constant equal to 2π.

6.4.2.2 Planetary Motion

As an example of a problem in which a universal, but not dimensionless, constant arises let us suppose that we wish to determine the form of the expression which specifies the time of revolution of one body round another under the action of gravitational forces. We assume that the two bodies are so far away from any other bodies that the only forces of significance are those of gravitational attraction between the two bodies themselves. One instance of this phenomenon is the movement of the earth round the sun, although the sun and the earth are not in fact so isolated that the effect of the gravitational attraction of other bodies can be entirely neglected. The phenomenon is also encountered in some problems of atomic structure where one body moves round another under the influence of attractive forces.

If each of the bodies is small compared with the distance between them, we might reasonably suppose that the time of revolution t depends only on the mass of each of the two bodies concerned (m_1, m_2) and on the mean distance r between them. That is, that the relation can be written in the form

$$t = \phi_1(m_1, m_2, r) \tag{6.6}$$

The function ϕ_1 expressed as a series would consist of terms of the form $km_1^a m_2^b r^c$ and the dimensional formula of each would be $[M^{a+b}L^c]$. Yet the left-hand side of eqn 6.6 has the dimensional formula $[T]$, and thus dimensional homogeneity is impossible whatever values might be given to a, b and c. This apparent deadlock suggests that the original list of relevant variables was incomplete. The missing magnitude is the gravitational constant G. It is relevant because the motion is governed by the force of attraction between the two bodies, and this is given by $F = Gm_1m_2/r^2$. Since, by convention, the magnitude of force is already defined by $F = ma$, the dimensional formula of F is $[MLT^{-2}]$ and so the dimensional formula of G is $[L^3M^{-1}T^{-2}]$.

We now start afresh with the relation

$$t = \phi_2(m_1, \; m_2, \; r, \; G) \tag{6.7}$$

The individual terms of the series by which ϕ_2 can be expressed are of the form $km_1{}^a m_2{}^b r^c G^d$ and the dimensional formula of each is $[M^a M^b L^c (L^3 M^{-1} T^{-2})^d]$ $\equiv [L^{c+3d} M^{a+b-d} T^{-2d}]$. If the dimensional formulae of the left- and right-hand sides of eqn 6.7 are now compared, we see at once that, for dimensional homogeneity in respect to time interval T, $1 = -2d$, that is $d = -\frac{1}{2}$. Dimensional homogeneity in respect to length requires $0 = c + 3d$, and so $c = 3/2$. The condition for dimensional homogeneity in respect to mass is $0 = a + b - d$ and, since $d = -\frac{1}{2}$, we obtain $a + b = -\frac{1}{2}$. However, no other equation involving a or b is available, and so definite values cannot be assigned to either a or b. All that can be done is to express one in terms of the other. If we choose to express b in terms of a, we obtain $b = -a - \frac{1}{2}$.

With the values deduced for b, c and d we can now write each term of the series in the form $km_1{}^a m_2{}^{-a-1/2} r^{3/2} G^{-1/2}$, i.e.

$$\frac{kr^{3/2}}{G^{1/2} m_2{}^{1/2}} \left(\frac{m_1}{m_2}\right)^a$$

The series as a whole is thus

$$\frac{r^{3/2}}{G^{1/2} m_2{}^{1/2}} \left\{ k_1 \left(\frac{m_1}{m_2}\right)^{a_1} + k_2 \left(\frac{m_1}{m_2}\right)^{a_2} + k_3 \left(\frac{m_1}{m_2}\right)^{a_3} + \ldots \right\}$$

and so the result may be expressed more briefly as

$$i = \frac{r^{3/2}}{G^{1/2} m_2{}^{1/2}} \phi_3 \left(\frac{m_1}{m_2}\right) \qquad (6.8)$$

Here too it is seen that, in each term of the series, the magnitude raised to an unknown power is itself simply a number (since it is the ratio of two like magnitudes). Thus the dimensional homogeneity of the equation is not affected by the value of the exponents a_1, a_2, etc. This of course is why applying the condition of dimensional homogeneity can tell us nothing about the values of these exponents.

A solution of this problem by normal mathematical methods is possible. It confirms eqn 6.8, and shows further that $\phi_3(m_1/m_2)$ is $[2\pi/\{(m_1/m_2) + 1\}]^{1/2}$. Although dimensional analysis did not yield the complete solution, it did give the important result that the period of revolution is proportional to the 3/2 power of the mean distance between the bodies.

6.4.2.3 'Friction' Loss for Flow in a Pipe

We turn now to a problem for which a complete solution is not given by normal algebraic methods. It is for such problems as this that dimensional analysis is especially valuable. Consider the drop in pressure occurring as a result of friction when a homogeneous fluid of constant density flows under steady conditions along a pipe of constant, circular, cross-sectional area. The general procedure is the same as in the previous examples. First, we list *all* the quantities which independently concern the phenomenon under investigation. Then we express their magnitudes in terms of the fundamental magnitudes. Finally we combine the magnitudes in such a way that the principle of dimensional homogeneity is satisfied.

The problem may be simplified to some extent at the start by realizing that, if 'end effects' are negligible, the drop in pressure Δp is proportional to the length

l of pipe over which it occurs. If an exactly similar length of the pipe were connected in series with the given length, then, if nothing else were altered, the drop in pressure in the extra length would equal that in the original length. Thus the pressure drop over the double length would be twice that obtained originally. We may therefore take as the quantity of primary interest the pressure drop per unit length, $\Delta p/l$. By taking $\Delta p/l$ as a single variable instead of considering Δp and l separately, the number of variables is reduced by one, and we shall be able to obtain more information from the analysis.

Unless the axis of the pipe is horizontal, gravity influences the flow along the pipe. However, for a cylinder of fluid of length l in a straight pipe of cross-sectional area A (see Fig. 6.1) the component of weight in the direction of flow is given by $\rho g A l \cos \alpha$, where ρ represents the (uniform) density of the fluid and

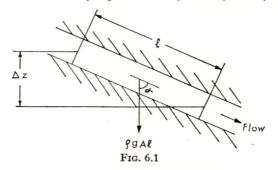

Fig. 6.1

α the angle between the forward flow direction and the downward vertical. For $l \cos \alpha$ we may write Δz, the difference in elevation between the two ends of the cylinder, and so the force per unit cross-sectional area caused by gravity is $\rho g \Delta z$. Thus, for a fluid of constant and uniform density, gravity forces may be accounted for simply by considering changes of 'piezometric pressure', $p^* = p + \rho g z$, rather than changes of pressure p alone. Here, for simplicity, we shall omit the asterisk from p^*, but remember that, if the pipe axis is not horizontal, piezometric pressure is considered instead of pressure.

We shall suppose for the moment that the roughness of the walls of the pipe is

TABLE 6.1

Quantity	Symbol	Dimensional formula
Drop in (piezometric) pressure per unit length of pipe	$\Delta p/l$	$\left[\dfrac{\text{Force/Area}}{\text{Length}}\right] \equiv \dfrac{[F/L^2]}{[L]}$ $\equiv \dfrac{[MLT^{-2}]}{[L^3]} \equiv [ML^{-2}T^{-2}]$
Diameter of pipe	d	$[L]$
Mean velocity of fluid (Rate of flow ÷ Area)	v	$[LT^{-1}]$
Absolute viscosity of fluid	μ	$[ML^{-1}T^{-1}]$
Density of fluid	ρ	$[ML^{-3}]$

not sufficient to affect the flow. The quantities involved in the problem may now be listed together with the corresponding dimensional formulae, see Table 6.1.

These are the only relevant quantities. If the pipe is horizontal the force tending to pull the fluid towards the earth is of no consequence: the case of a pipe which is not horizontal is taken care of by considering piezometric pressure. Surface tension acts only where there is a surface in contact with another fluid, and so is of no account in a pipe full of a single fluid.

The quantity of primary interest is $\Delta p/l$ and so we seek a relation of the form
$$\Delta p/l = \phi(d, v, \mu, \rho)$$
In general the function is expressible as a series of terms each of the form $k d^a v^b \mu^c \rho^e$†, but as all such terms must have the same dimensional formula it is sufficient here to consider one term only. For dimensional homogeneity
$$[\Delta p/l] \equiv [d^a v^b \mu^c \rho^e] \tag{6.9}$$
i.e.
$$[ML^{-2}T^{-2}] \equiv [L]^a [LT^{-1}]^b [ML^{-1}T^{-1}]^c [ML^{-3}]^e$$
$$\equiv [L]^{a+b-c-3e} [M]^{c+e} [T]^{-b-c}$$

The powers of the fundamental magnitudes [L], [M] and [T] on the two sides of this relation must individually be identical. Three separate equations are therefore obtained.

Equating exponents of [L]: $-2 = a + b - c - 3e$
Equating exponents of [M]: $1 = c + e$
Equating exponents of [T]: $-2 = -b - c$

With four unknowns but only three equations a complete solution is impossible. It is, however, possible to determine three of the exponents in terms of the remaining one. The choice of the exponent to remain unknown is quite arbitrary, but here let us select c. From the second and third equations we obtain $e = 1 - c$ and $b = 2 - c$; these values substituted in the first then give $a = -1 - c$. The relation 6.9 may now be written
$$[\Delta p/l] \equiv [d^{-1-c} v^{2-c} \mu^c \rho^{1-c}] \equiv \left[\rho v^2 d^{-1} \left(\frac{\mu}{vd\rho}\right)^c\right]$$

The group $\mu/vd\rho$ is dimensionless, and this is a necessary result because no restriction was placed on the value of the exponent c. Any value of c is consistent with the requirement of dimensional homogeneity. The group $\mu/vd\rho$ is the reciprocal of the Reynolds number, Re, for a circular pipe (Section 7.4), and the expression may thus be written
$$[\Delta p/l] \equiv [\rho v^2 d^{-1} (Re)^{-c}]$$
The series of terms representing the desired function is of the form
$$\Delta p/l = k_1 \rho v^2 d^{-1} (Re)^{-c_1} + k_2 \rho v^2 d^{-1} (Re)^{-c_2} + \ldots$$
$$= \rho v^2 d^{-1} \{k_1 (Re)^{-c_1} + k_2 (Re)^{-c_2} + \ldots\}$$
$$= \rho v^2 d^{-1} \phi(Re) \tag{6.10}$$

The pressure drop per unit length depends, then, in some unspecified manner on the Reynolds number. Equation 6.10 may be rearranged to give
$$\frac{(\Delta p)d}{l \rho v^2} = \phi(Re) \tag{6.11}$$

† To avoid confusion with d = diameter the symbol d is not used as an exponent.

The left-hand side of eqn 6.11 constitutes a group which, like Reynolds number, is dimensionless. The original problem in which there were five variables (one dependent, four independent) has now been brought down to one with only two parameters, one a function of the other. A considerable simplification has thus been achieved. Experimental results can be represented on a single graph in which $(\Delta p)d/l\rho v^2$ is plotted against Reynolds number. (In fact, using Darcy's formula for the head lost to friction, $h_f = (4fl/d)(v^2/2g)$, we see that

$$\frac{(\Delta p)d}{l\rho v^2} = \frac{\rho g h_f d}{l\rho v^2} = \frac{\rho g d}{l\rho v^2}\left(\frac{4fl}{d}\frac{v^2}{2g}\right) = 2f$$

Therefore the curve would be similar to the graph of friction factor f against Reynolds number for smooth pipes—as found in textbooks on the mechanics of fluids.)

Dimensional analysis provides no information about whether the flow is laminar or turbulent. For laminar flow, Poiseuille's formula gives

$$\Delta p = 128\mu l Q/\pi d^4$$

and so the left-hand side of eqn 6.11 is then

$$\frac{128\mu l Q}{\pi d^4}\frac{d}{l\rho v^2} = \frac{128\mu l}{\pi d^4}\frac{\pi}{4}d^2 v\frac{d}{l\rho v^2} = 32\frac{\mu}{v d \rho} = 32(Re)^{-1}$$

In this example, which particular dimensionless parameters are finally obtained depends on which exponent was selected to remain undetermined. In the preceding analysis the equations were solved to give a, b and e in terms of c. But if, say, b, c and e had been found in terms of a the result would have been

$$\frac{\Delta p}{l} = \frac{\rho^2 v^3}{\mu}\phi(Re), \quad \text{i.e.} \quad \frac{(\Delta p)\mu}{l\rho^2 v^3} = \phi(Re) \tag{6.12}$$

This result is just as correct as the other. It merely presents the variables in a different way. Again experimental results could be depicted on a single graph: this time $(\Delta p)\mu/l\rho^2 v^3$ would be plotted against Reynolds number. The resulting curve would have a different appearance but would be just as valid as the previous one. The choice of dimensionless parameters is partly a matter of convention; it depends too on the use to be made of the results. For example, if the effect of viscosity is to be studied, eqn 6.12 would be rather inconvenient in use because μ appears both in $(\Delta p)\mu/l\rho^2 v^3$ and in Re: for this purpose eqn 6.11, in which only Re incorporates μ, would be better.

Although such alternative forms of the solution are equally logical, it must be remembered that a result obtained by dimensional analysis can be no more correct than the initial formulation of the problem. Care is always necessary in compiling the list of quantities which are directly relevant to the situation.

If, in the problem just discussed, the roughness of the walls of the pipe is sufficient to affect the flow, another variable is introduced—for example, the average height k of the 'bumps' on the surface. An additional exponent therefore enters the equations which express the conditions for homogeneity. The number of fundamental magnitudes and thus the number of equations remain unchanged,

§6.4]   DIMENSIONAL ANALYSIS   63

however, and so a second exponent must remain undetermined. Corresponding with eqn. 6.10, one form of the final result would be

$$\Delta p/l = k_1 \rho v^2 d^{-1}(Re)^{-c_1}(k/d)^{f_1} + k_2 \rho v^2 d^{-1}(Re)^{-c_2}(k/d)^{f_2} + \ldots$$
$$= \rho v^2 d^{-1}\{k_1(Re)^{-c_1}(k/d)^{f_1} + k_2(Re)^{-c_2}(k/d)^{f_2} + \ldots\}$$
$$= \rho v^2 d^{-1}\phi(Re, k/d)$$

Expressing the unknown function in the form of a series is an essential part of this method of dimensional analysis. Because the dimensional formula of a single term of that series is examined there is a temptation to assume that the series consists, in fact, of that term only. This false assumption leads to the error of supposing that the dependent variable in any problem may be equated to a product of a power of each of the independent variables together with a constant numerical coefficient. Although such a procedure may be justified in a few particular cases, it is not valid in general.

6.4.3   THE 'PI' THEOREM

The method of dimensional analysis described in the previous section is often known as the indicial method or Rayleigh's method after Lord Rayleigh (1842–1919) who was probably the first to use such arguments extensively. We now turn attention to a somewhat different technique in which the so-called 'Pi' theorem is used. The first explicit statement of the theorem was made by Edgar Buckingham (1867–1940), although some earlier workers had used equivalent results. The method based on the theorem is therefore sometimes known as Buckingham's method and occasionally as the General method.

The Pi theorem is not a means of extending dimensional analysis to cover problems which cannot be tackled by Rayleigh's method. It does, however, form the basis of a logical procedure of analysis and enables results of an investigation to be presented in the most general form.

The 'Pi' of the title refers to the Greek capital letter Π which is used as a symbol for 'product'. (There is no connection with $\pi = 3.14\ldots$.) The object of Buckingham's method is to assemble all the magnitudes entering the problem into a number of dimensionless products (Πs), and the desired relation connecting individual magnitudes is then an algebraic expression relating the Πs.

A rigorous proof of the theorem requires the theory of simultaneous linear equations and this need not detain us here. It is helpful, however, to consider the result obtained by Rayleigh's method in eqn 6.11. In the problem examined there we began with *five* magnitudes treated as variables ($\Delta p/l, d, v, \mu, \rho$). The dimensional formulae of these magnitudes could be expressed in terms of powers of *three* fundamental magnitudes (L, M, T) and thus three simultaneous equations could be used for determining—as far as possible—the exponents. The final result was a relation between $5 - 3 = 2$ dimensionless parameters. If there had originally been six variables and still only three fundamental magnitudes, one more exponent would have remained arbitrary and the final expression would have related $6 - 3 = 3$ dimensionless parameters.

These results provide an illustration of the Pi theorem, which may in essence be stated thus:

Suppose that the magnitude Q_1 of some physical quantity depends on other,

independent, magnitudes Q_2, Q_3, etc. (up to Q_n) and on no others. Then the general relationship may be written
$$Q_1 = \phi(Q_2, Q_3, \ldots, Q_n)$$
It is always possible to rearrange any equation so that it has zero on one side, and so an alternative way of writing the relation is
$$\phi_1(Q_1, Q_2, Q_3, \ldots, Q_n) = 0 \qquad (6.13)$$
(The form of the function ϕ_1 is of course different from that of ϕ.)

This equation must be dimensionally homogeneous. Consequently, dimensional constants must be included if necessary. Furthermore, there should be no relation other than this one connecting any of the magnitudes; that is, the magnitudes must be otherwise independent.

The Pi theorem now states that if m is the number of distinct fundamental magnitudes required to express the dimensional formulae of all the n magnitudes, then these n magnitudes may be grouped into $n - m$ independent dimensionless Π terms, and the relation may be written
$$\phi(\Pi_1, \Pi_2, \ldots, \Pi_{n-m}) = 0 \qquad (6.14)$$
(The implications of the word 'distinct' in the above statement of the theorem will be examined later.)

A comparison of eqns 6.13 and 6.14 shows the advantage of dimensional analysis from the experimenter's point of view: the number of variables needing separate consideration is reduced from n to $n - m$.

We must now consider how these dimensionless products, the Πs, are to be formed. The simplest way of ensuring that they are independent (for example, one Π must not be merely the square of another Π or its reciprocal) is that devised by Buckingham.

From among the n original variables m of them are selected to form a *recurring set* (sometimes called a 'pro-basic set'). The choice of the m variables for this purpose is not entirely free: for one thing, it is essential that the recurring set shall involve all the m fundamental magnitudes. For example, in a problem in mechanics the fundamental magnitudes are usually those of length, mass and time interval, and so $m = 3$. Of the three variables chosen as the recurring set at least one must involve length in its dimensional formula, at least one must involve mass, and at least one must involve time interval. When the recurring set has been chosen, each of the remaining $n - m$ magnitudes is in turn put with those of the recurring set in such a way as to form a dimensionless group, that is a Π. It is clear, however, that if variables in the recurring set can *by themselves* be formed into a dimensionless group, they cannot form a dimensionless group in association with another variable which is not (in general) dimensionless.

It is now evident why all the fundamental magnitudes must be involved at least once in the recurring set. If a group is to be dimensionless the components of its dimensional formula must 'cancel'. The recurring set must therefore provide the opportunity for this 'cancelling' of any dimensional components which are involved in the other magnitudes. For example, if the recurring set did not involve at all the fundamental magnitude of mass, then it would be impossible to form a dimensionless product with another magnitude whose dimensional formula did include mass.

§6.4]   DIMENSIONAL ANALYSIS   65

The necessary features of the recurring set, then, are these. Its constituent magnitudes must among themselves involve each of the fundamental magnitudes at least once, and yet it must not be possible to form a dimensionless group from the recurring set alone.

To illustrate the application of the theorem we may take the example of fluid flow through a smooth pipe of circular cross-section—a problem which has already been examined by Rayleigh's method. There are five variables, the dimensional formulae of which may be expressed in terms of three fundamental magnitudes (i.e. $n = 5$, $m = 3$) and so $5 - 3 = 2$ dimensionless products are to be expected. For the recurring set we require three of the five variables which together involve [L], [M] and [T]. As the list in Section 6.4.2.3 shows, the trio, v, μ, ρ is suitable. Each Π is now formed from the recurring set and one of the remaining magnitudes. One Π is therefore constructed from v, μ, ρ and $\Delta p/l$; the other from v, μ, ρ and d.

For the group v, μ, ρ, $\Delta p/l$ let the dimensionless arrangement take the form $v^{a_1} \mu^{b_1} \rho^{c_1} \Delta p/l$. (The exponent of $\Delta p/l$ has been chosen as unity: any one exponent in a dimensionless product may be made unity by taking the appropriate root of the product, and the result remains dimensionless.) The dimensional formula of this product is therefore

$$[LT^{-1}]^{a_1}[ML^{-1}T^{-1}]^{b_1}[ML^{-3}]^{c_1}[ML^{-2}T^{-2}] \equiv [L^0 M^0 T^0]$$

Balancing the exponents of [L], [M] and [T] we get, respectively,

$$a_1 - b_1 - 3c_1 - 2 = 0$$
$$b_1 + c_1 + 1 = 0$$
$$-a_1 - b_1 - 2 = 0$$

from which $a_1 = -3$, $b_1 = 1$, $c_1 = -2$. Therefore the product Π_1 may be written $v^{-3} \mu \rho^{-2} \Delta p/l$.

Similarly, let Π_2 take the form $v^{a_2} \mu^{b_2} \rho^{c_2} d$. Again balancing the exponents of the fundamental magnitudes [L], [M], [T], we arrive at $a_2 = 1$, $b_2 = -1$, $c_2 = 1$, and Π_2 may then be written $v \mu^{-1} \rho d$.† The relation between Π_1 and Π_2 may be put in the form

$$\phi(\Pi_1, \Pi_2) = 0, \quad \text{i.e.} \quad \phi\left(\frac{\mu \Delta p/l}{v^3 \rho^2}, \frac{v d \rho}{\mu}\right) = 0 \quad (6.15)$$

If the variable of primary interest is $\Delta p/l$, eqn 6.15 may be rearranged to separate the Π containing it:

$$\frac{\mu \Delta p/l}{v^3 \rho^2} = \phi_1\left(\frac{v d \rho}{\mu}\right)$$

or

$$\frac{\Delta p}{l} = \frac{v^3 \rho^2}{\mu} \phi_1\left(\frac{v d \rho}{\mu}\right) = \frac{v^3 \rho^2}{\mu} \phi_1(Re) \quad (6.16)$$

This rearrangement is entirely justifiable. The equation 'some function of x and $y = 0$' may always be transformed by normal algebraic processes to give 'x = some (other) function of y'.

It will be noticed that eqn 6.16 is identical with eqn 6.12.

The reader, having followed the analysis of this problem both by Rayleigh's

† An alternative method of constructing the Πs is mentioned later in this Section.

method and by Buckingham's, may feel that the latter offers no advantages over Rayleigh's. In the example just considered, Rayleigh's method does in fact probably score in simplicity. Buckingham's method comes particularly into its own in more complex problems. An instructive example is one which Buckingham himself studied. In the light of this example it will be easier to answer some other questions, and so we consider now a screw propeller operating in a fluid of constant density such as sea water.

The quantity of particular interest is the thrust produced by the propeller, and our aim is to discover how the magnitude of this force depends on the magnitudes of other quantities entering the problem.

The fluid is assumed to be homogeneous: if, for example, sea water is considered it is assumed to contain no bubbles of air or water vapour. (That is, 'cavitation' is not occurring.) Also, the propeller is assumed to be remote from all other surfaces. This assumption is not justified for a ship's propeller, the action of which is in fact greatly modified by the proximity of the hull. If we wished to include the effect of the hull in modifying the flow through and around the propeller, then, even for a given shape of hull and a given position of the propeller relative to the hull, some suitable length representing the distance between hull and propeller would have to be included in our list of relevant quantities.

For our present purpose, however, we leave these complications aside. First, all the relevant quantities are listed, as in Table 6.2.

TABLE 6.2

Quantity	Symbol	Dimensional formula
Thrust (force)	F	$[MLT^{-2}]$
Density of fluid	ρ	$[ML^{-3}]$
Diameter of propeller	D	$[L]$
Linear speed (i.e. 'speed of advance')	v	$[LT^{-1}]$
Number of revolutions in unit time	N	$[T^{-1}]$
Viscosity of fluid	μ	$[ML^{-1}T^{-1}]$
Weight per unit mass	g	$[LT^{-2}]$

The last quantity is included because, unless the propeller is very deeply immersed, it causes disturbances on the surface of the liquid; in other words, work is done against gravity in moving the surface vertically, and so the weight of the fluid enters the problem. Unless our concern is with tiny, toy propellers, the effects of surface tension may be neglected. Thus the number of relevant quantities is 7, the number of fundamental magnitudes is 3 (L, M, T), and so $7 - 3 = 4$ dimensionless products are to be expected. That is, the relation may be expressed in the form

$$\phi(\Pi_1, \Pi_2, \Pi_3, \Pi_4) = 0$$

From the seven variables we select three (because the number of fundamental magnitudes is three) for use as a recurring set. The trio F, ρ, D is suitable. (Note that the set D, v, N, for example, is not suitable because it nowhere includes the fundamental magnitude [M]. Another unsuitable set is F, ρ, μ, even though it involves all the fundamental magnitudes, because a dimensionless group—$F\rho\mu^{-2}$—can be formed from these variables alone. Attempts to use F, ρ, μ as a recurring set would yield inconsistent, and therefore insoluble, equations for the exponents in the Πs.)

With F, ρ, D as the recurring set the dimensionless Πs take the forms

$$F^{a_1}\rho^{b_1}D^{c_1}v; \quad F^{a_2}\rho^{b_2}D^{c_2}N; \quad F^{a_3}\rho^{b_3}D^{c_3}\mu; \quad F^{a_4}\rho^{b_4}D^{c_4}g$$

The exponents are readily determined. For example, the dimensional formula of the first Π is

$$[MLT^{-2}]^{a_1}[ML^{-3}]^{b_1}[L]^{c_1}[LT^{-1}]$$

For this to be dimensionless

$$\begin{aligned} a_1 - 3b_1 + c_1 + 1 &= 0 \quad \text{(condition on L)} \\ a_1 + b_1 &= 0 \quad \text{(condition on M)} \\ -2a_1 - 1 &= 0 \quad \text{(condition on T)} \end{aligned}$$

from which $a_1 = -\tfrac{1}{2}$, $b_1 = \tfrac{1}{2}$, $c_1 = 1$.

The first product may therefore be written $F^{-1/2}\rho^{1/2}Dv$. The rather inconvenient fractional exponents may be removed by squaring the expression:

$$\Pi_1 = \rho D^2 v^2 / F$$

In a similar way the other Πs are determined:

$$\Pi_2 = \rho D^4 N^2 / F; \quad \Pi_3 = \mu^2 / F\rho; \quad \Pi_4 = \rho D^3 g / F$$

The required relation may thus be expressed in the form

$$\phi\left(\frac{\rho D^2 v^2}{F}, \frac{\rho D^4 N^2}{F}, \frac{\mu^2}{F\rho}, \frac{\rho D^3 g}{F}\right) = 0 \tag{6.17}$$

and we conclude that any correct and 'complete' equation relating the seven variables may be reduced to this form.

The recurring set F, ρ, D is not the only possible choice. In fact, several other sets are suitable, and analyses performed with them would yield equations looking different from eqn 6.17 but essentially the same. For example, if the set ρ, D, v were chosen the result would be

$$\phi\left(\frac{\rho D^2 v^2}{F}, \frac{DN}{v}, \frac{\rho Dv}{\mu}, \frac{Dg}{v^2}\right) = 0 \tag{6.18}$$

Only the first of the arguments in the function in eqn 6.18 corresponds directly with any in eqn 6.17. The new dimensionless products are not, however, independent of those in eqn 6.17 but may be obtained from them. For instance, from the first two products in eqn 6.17,

$$\frac{\Pi_1}{\Pi_2} = \frac{\rho D^2 v^2}{F} \frac{F}{\rho D^4 N^2} = \frac{v^2}{D^2 N^2} = \left(\frac{DN}{v}\right)^{-2}$$

which is a power of the second Π in eqn 6.18. The other two 'new' products may be obtained similarly from a combination of the original ones.

This sort of transformation of Πs is often useful in any case. It must be remembered, however, that the number of independent Πs must always be that specified by the Pi theorem, $n - m$: multiplying or dividing one Π by another does not reduce the number of independent Πs necessary in the expression. In the present example, for instance, no matter what manipulation is done four distinct Πs must remain.

Care must be taken that these four Πs are in fact distinct, that is, independent of each other. For instance, from the result $\phi(\Pi_1, \Pi_2, \Pi_3, \Pi_4) = 0$ we could combine Π_1 and Π_2 in the form Π_1/Π_2, but if this new Π were substituted for Π_3 we should obtain $\phi(\Pi_1, \Pi_2, \Pi_1/\Pi_2, \Pi_4) = 0$. In this modified expression the third Π is not independent of the first two; thus there are only three *independent* Πs—and the phenomenon under investigation has been made to appear entirely uninfluenced by the original Π_3. (Substituting Π_1/Π_2 for Π_1, say, would of course be acceptable.) Evidently, then, any new Π which is obtained by manipulation of others may be substituted only for one of those used in that manipulation.

If the magnitude of one of the original quantities is to be regarded as the dependent variable, then that magnitude must appear in only one of the dimensionless products. This may be ensured by excluding that magnitude from the recurring set. For example, if the thrust F is to be taken as the dependent variable, then the recurring set must be one not including F. The analysis leading to eqn 6.18 meets this requirement and the result may be rearranged

$$\frac{F}{\rho D^2 v^2} = \phi_1\left(\frac{DN}{v}, \frac{\rho D v}{\mu}, \frac{Dg}{v^2}\right)$$

or

$$F = \rho D^2 v^2 \phi_1\left(\frac{DN}{v}, \frac{\rho D v}{\mu}, \frac{Dg}{v^2}\right)$$

All these results could have been obtained by Rayleigh's method but the process would have been long and tedious. Once the recurring set has been chosen Buckingham's technique is straightforward and rapid since the practised user of the method can usually insert the exponents of the magnitudes in the Πs by inspection.

An alternative method of constructing the Πs involves expressing the usual fundamental magnitudes in terms of the dimensional formulae of the members of the recurring set. Consider the recurring set ρ, D, v in the present example. Then, since $[\rho] \equiv [ML^{-3}]$ and $[D] \equiv [L]$,

$$[M] \equiv [\rho L^3] \equiv [\rho D^3].$$

Also, since $\quad [v] \equiv [LT^{-1}]$,

$$[T] \equiv [Lv^{-1}] \equiv [Dv^{-1}].$$

Therefore

$$[\mu] \equiv [ML^{-1}T^{-1}] \equiv [\rho D^3 D^{-1} D^{-1} v] \equiv [\rho D v].$$

As μ thus has the same dimensional formula as ρDv, the ratio $\rho Dv/\mu$ is dimensionless, and this is the Π containing μ in eqn 6.18. The Π containing g can be similarly determined:

$$[g] \equiv [LT^{-2}] \equiv [D(Dv^{-1})^{-2}] \equiv [D^{-1}v^2]$$

and so $g/(D^{-1}v^2) = Dg/v^2$ is dimensionless.

In the statement of the Pi theorem at the beginning of this section it was said that m represents the number of *distinct* fundamental magnitudes. What does the word 'distinct' imply?

Usually no special thought is needed to ensure that the fundamental magnitudes are distinct. Occasionally, however, the condition for homogeneity in respect to one of the fundamental magnitudes proves to be no different from the condition for homogeneity in respect to another. The number of *different* equations used to determine exponents is thus reduced by one. In Rayleigh's method this state of affairs is evident as soon as the equations are written down; one more exponent than expected remains arbitrary, and in the final result the number of dimensionless products is one greater than at first anticipated. (One instance in which two homogeneity conditions coincide is in the first example discussed in Section 6.5.) Occasionally too it may happen that the equation given by the homogeneity condition for one fundamental magnitude is the same as that obtained by adding (or subtracting) the equations for the homogeneity of two other fundamental magnitudes.

When the Pi theorem is used in such a problem it is found either that it is impossible, with m variables, to choose a recurring set which does not itself form a dimensionless group, or that it is impossible to determine completely the exponents of the magnitudes forming the Πs. It is then necessary to use a recurring set in which the number of magnitudes is one smaller than in the first attempt. Again the final result has one more dimensionless product than at first expected.

For example, the condition for homogeneity in respect to mass may be identical with that for homogeneity in respect to time interval. This simply indicates that it would have been sufficient to regard [M/T] (corresponding to rate of increase of mass) as a single fundamental magnitude instead of [M] and [T] separately.† In short, the number of Πs equals the number of quantities entering the problem minus the *maximum* number whose magnitudes will not by themselves form a Π (and are therefore suitable for use as a recurring set).

6.4.3.1 Reducing the Number of Πs

The greater the number m of distinct fundamental magnitudes the smaller is the number $n - m$ of Πs and the more informative is the final result. Sometimes it is possible to use a greater number of distinct fundamental magnitudes than usual and thus to increase the precision of the result. An interesting example of this dodge arises when we consider the vertical fall of a solid sphere with a small steady velocity through an infinite expanse of a homogeneous viscous fluid.

A small solid sphere, falling through the fluid under its own weight, accelerates until the net downward force on it is zero. No further acceleration is then possible and the sphere is said to have reached its 'terminal velocity'. Under these

---

† Identical conditions for homogeneity in respect to two fundamental magnitudes perhaps arise most commonly in the investigation of systems of static forces. In such systems the only quantities involved are lengths, areas, volumes and forces. Consequently, only two fundamental magnitudes—those of length and force—are required, and there are only two separate conditions for homogeneity. However, if [L], [M] and [T] are taken as fundamental magnitudes, homogeneity conditions are obtained for each, but the conditions in respect to [M] and [T] are essentially the same. In analysing a system involving only constant, static forces, time can be saved by using [L] and [F] as fundamental magnitudes rather than [L], [M] and [T].

conditions, the viscous forces resisting the motion are exactly counterbalanced by the weight of the sphere minus its buoyancy. We wish to discover how the 'terminal velocity' depends on other features of the situation.

We now assume not only that the motion of the sphere has become completely steady, but also that the forces required to accelerate individual particles of fluid around the sphere are negligible compared with the viscous forces. (This condition is met if the Reynolds number of the flow is exceedingly small.) Here, then, is the unusual circumstance that the relation Force = Mass × Acceleration does not enter a problem of motion. As the defining equation $F = ma$ is not used, the magnitude of force may be separately defined, and so in this instance it is possible to regard the magnitude of force as independent of those of mass and acceleration. Four fundamental magnitudes—[L], [M], [T] and [F]—may now be employed.

There are six physical quantities involved as shown in Table 6.3.

TABLE 6.3

Quantity	Symbol	Dimensional formula
Terminal velocity of sphere	v	$[LT^{-1}]$
Diameter of sphere	d	$[L]$
Density of fluid	ρ	$[ML^{-3}]$
Density of sphere	ρ_s	$[ML^{-3}]$
Viscosity of fluid	μ	$[FL^{-2}T]$
Weight per unit mass	g	$[FM^{-1}]$

It is important to notice that the dimensional formula of viscosity is not written $[ML^{-1}T^{-1}]$ here. The form $[FL^{-2}T]$ is arrived at directly from the definition of the magnitude of viscosity as (stress) ÷ (velocity gradient), i.e. $[FL^{-2}] \div [LT^{-1}/L]$. The more usual dimensional formula is derived from this one by using the relation $[F] \equiv [\text{mass}][\text{acceleration}] \equiv [MLT^{-2}]$ which we wish to avoid in this analysis. The usual dimensional formula $[LT^{-2}]$ for g is likewise based on the relation $F = ma$, and so the direct form $[FM^{-1}]$ is used here.

With six relevant quantities and four fundamental magnitudes we may expect two dimensionless products. One of these is evident by inspection: the ratio of the densities ρ_s/ρ. The other may be written $d^a \rho^b \mu^c g^e v$ and the exponents readily determined as $a = -2, b = -1, c = 1, e = -1$. The result is thus

$$\phi\left(\frac{\rho_s}{\rho}, \frac{v\mu}{d^2\rho g}\right) = 0$$

or

$$v = \frac{d^2 \rho g}{\mu} \phi_1\left(\frac{\rho_s}{\rho}\right) \tag{6.19}$$

Had only three fundamental magnitudes been used, say [L], [M] and [T], then $6 - 3 = 3$ dimensionless products would have been obtained instead of two. A possible form of the result would be

$$v = \frac{\mu}{\rho d} \phi\left(\frac{\rho^2 d^3 g}{\mu^2}, \frac{\rho_s}{\rho}\right)$$

§6.4] DIMENSIONAL ANALYSIS 71

The greater precision achieved by using the extra fundamental magnitude is evident. The same improvement could, indeed, be effected in another, still simpler, way. Mass is a measure of the quantity of matter, and matter has two fundamental properties—inertia and gravitational attraction. Inertia, that is 'reluctance to be accelerated', does not enter the present problem, and the only gravitational attractions of significance are that between the sphere and the earth, and that between the fluid and the earth. In other words, mass figures in the problem only in determining weights. Mass may therefore be omitted as a quantity with a fundamental magnitude in favour of weight (i.e. force), and so the list of relevant quantities becomes as shown in Table 6.4.

TABLE 6.4

Quantity	Symbol	Dimensional formula
Terminal velocity	v	$[LT^{-1}]$
Diameter of sphere	d	$[L]$
Specific weight (i.e. weight per unit volume) of fluid	w	$[FL^{-3}]$
Specific weight of sphere	w_s	$[FL^{-3}]$
Viscosity of fluid	μ	$[FL^{-2}T]$

The weight per unit mass no longer need be included in the list, although if it were included its exponent would be found to be zero because it would be the only magnitude involving that of mass. Here there are five variables and three fundamental magnitudes and so two Πs may be expected in the result. To correspond with eqn. 6.19 we obtain

$$v = \frac{d^2 w}{\mu} \phi\left(\frac{w_s}{w}\right) \tag{6.20}$$

Since

$$\frac{\text{weight}}{\text{volume}} = \frac{\text{mass}}{\text{volume}} \times \frac{\text{weight}}{\text{mass}}$$

$w = \rho g$ and eqns 6.19 and 6.20 are seen to be identical.

The greater precision obtainable when the number of independent fundamental magnitudes can be increased is very valuable. Unfortunately, the attractions of a more refined result have lured some writers into a fallacy, which, because it has acquired some currency, is mentioned briefly here. Instead of regarding length as having a single fundamental magnitude $[L]$, these writers have supposed that lengths in different directions—for example, L_x, L_y, L_z, respectively parallel to rectangular coordinate axes—may have magnitudes independent of each other. Dimensional homogeneity has then been sought with respect to the magnitudes of each of these component lengths individually.

To obtain the 'right' answers to particular problems (that is, answers already known from other information), those attempting this more detailed analysis have often been obliged to adopt arguments notable more for ingenuity than for irreproachable logic. Moreover, those more precise results claimed for the method fall, it seems, into one of only two categories: (1) those which are wrong (perhaps only partly wrong in that they describe particular rather than general conditions), or (2) those obtainable by other techniques for minimizing the number of Πs.

It is not possible in a paragraph or two to refute the fallacy rigorously. It is therefore simply recorded here that nothing, except complexity and confusion, is achieved by regarding lengths in different directions as having separate fundamental magnitudes.

6.5 Dimensional Formula of Plane Angle

For nearly all quantities, convention has determined how the magnitudes shall be defined, and discussion on the matter is, for practical purposes, now closed. But two quite separate definitions of the magnitude of plane angle are in common use. One of these definitions makes no reference to the magnitude of any other quantity, and thus defines the magnitude of plane angle fundamentally. This definition provides the scale of degrees. A degree is defined simply as 1/360 of a complete turn: the magnitude of a particular angle can be specified as a certain number of degrees (or as a certain fraction of a complete turn) without using the concept of length, or of any other quantity, and certainly without using any other magnitude. The magnitude of an angle measured by a scale of degrees is therefore a fundamental one, and the corresponding dimensional formula could be written [A].

The other definition of the magnitude of plane angle treats it as derived. The magnitude is then defined as the ratio which the length of a circular arc, subtending the angle, bears to the radius of the arc. If the length of the arc equals the radius, the magnitude of the angle is said to be one radian. Since, by this definition, the magnitude of the angle is the ratio of two lengths, the corresponding dimensional formula is $[L] \div [L] \equiv [1]$; in other words, the magnitude is dimensionless.

Which dimensional formula should be used in dimensional analysis? [A] or [1]? Since dimensional analysis cannot yield any information about the way *numbers* enter a relation (except those numbers which are exponents in dimensional formulae) the answer to the question seems obvious: by using [A] we should get some information about the influence of angle on a phenomenon, whereas by regarding the magnitude of angle as dimensionless no such information is obtainable. In fact, the answer is not quite so simple as this.

In some instances there is distinct advantage in using [A]. Let us consider, for example, the twisting of an elastic wire of circular cross-section. The torque T required may be supposed to depend on θ the angle of twist, d the diameter of the wire, l its length, and G its modulus of rigidity. Now the modulus of rigidity is defined by the relation $\tau = G\alpha$, where τ represents the uniform shear stress exerted over the top of a parallelepiped of the substance, and α represents the resulting angular displacement of the sides. Let us first regard the magnitude of the angle α as dimensionless. Then G has the same dimensional formula as the stress τ, that is $[ML^{-1}T^{-2}]$.

Considering a typical term of the series which expresses the function of θ, d, l and G, we have
$$T \equiv [\theta^a d^b l^c G^e]$$

§6.5] DIMENSIONAL FORMULA OF PLANE ANGLE

If only the magnitudes of length, mass and time interval are taken as fundamental, the relation connecting the dimensional formulae becomes
$$[ML^2T^{-2}] \equiv [1]^a[L]^b[L]^c[ML^{-1}T^{-2}]^e$$
The conditions for homogeneity in respect to [M] and [T] both yield $e = 1$. From the condition for homogeneity in respect to [L] we have
$$2 = b + c - e$$
whence $b + c = 3$, i.e. $b = 3 - c$. Thus the formula for the torque required to twist the wire must have the form
$$T = k_1\theta^{a_1}d^{3-c_1}l^{c_1}G + k_2\theta^{a_2}d^{3-c_2}l^{c_2}G + \ldots$$
$$= Gd^3\left\{k_1\theta^{a_1}\left(\frac{l}{d}\right)^{c_1} + k_2\theta^{a_2}\left(\frac{l}{d}\right)^{c_2} + \ldots\right\}$$
$$= Gd^3\phi\left(\theta, \frac{l}{d}\right) \tag{6.21}$$

Now let us admit angle as a quantity having a fundamental magnitude [A]. The dimensional formula of G is that of τ/α, and this now becomes
$$[ML^{-1}T^{-2}A^{-1}].$$
Again dimensional homogeneity requires
$$T \equiv [\theta^a d^b l^c G^e]$$
but in more detailed form this is now
$$[ML^2T^{-2}] \equiv [A]^a[L]^b[L]^c[ML^{-1}T^{-2}A^{-1}]^e$$
The conditions for dimensional homogeneity in respect to [M] and [T] again give $e = 1$. Homogeneity in [A] requires $0 = a - e$, and thus $a = 1$. As before, the condition on [L] gives $b + c = 3$, and thus $b = 3 - c$. We can now say that the expression must take the form
$$T = k_1\theta d^{3-c_1}l^{c_1}G + k_2\theta d^{3-c_2}l^{c_2}G + \ldots$$
$$= \theta Gd^3\left\{k_1\left(\frac{l}{d}\right)^{c_1} + k_2\left(\frac{l}{d}\right)^{c_2} + \ldots\right\}$$
$$= \theta Gd^3\phi(l/d) \tag{6.22}$$

This result is clearly more informative than eqn 6.21, for it tells us that the required torque is directly proportional to the angle of twist.† So long as the magnitude of angle is regarded as dimensionless, dimensional analysis can offer no information at all about the effect of an angle on a phenomenon.

In the foregoing example the influence of angle is direct. Usually, however, angle has an indirect effect. That is, angle enters a relation only in determining trigonometrical functions such as sin, cos and tan, and these, being by definition ratios of lengths, are dimensionless. The results then do not depend on the scale used for measuring the angle: for example, the magnitude of the sine of one-third of a right angle is 0·5 whether the angle is expressed as 30° or as $\pi/6$ radians.

Dimensional analysis can yield information only about the *direct* effect of

---

† The algebraic solution is $T = \frac{\pi}{32}\theta Gd^3\left(\frac{l}{d}\right)^{-1} = \frac{\pi}{32}\theta Gd^4/l$.

UDA-F

angle. As an illustration, let us consider again the example of the simple pendulum already discussed in Section 6.4.2.1. We again suppose that the period of oscillation t is a function of the length l of the pendulum, the mass m of the bob, the weight per unit mass g, and the amplitude α of the swing on each side of the vertical. By considering the dimensional formula of a typical term in the series expressing the function we have

$$[t] \equiv [l^a m^b g^c \alpha^d] \qquad (6.23)$$

and expressing the magnitudes in terms of the fundamental ones of length, mass, time interval *and angle* we obtain

$$[T] \equiv [L]^a [M]^b [LT^{-2}]^c [A]^d$$

As before, dimensional homogeneity in respect to [L], [M] and [T] requires $a = \frac{1}{2}$, $b = 0$, $c = -\frac{1}{2}$. Dimensional homogeneity in respect to the magnitude of angle requires $d = 0$. From this we might deduce that the angle of swing has no effect whatever on the period of oscillation.

However, an exact algebraic solution of the problem gives

$$t = 2\pi \left(\frac{l}{g}\right)^{1/2} \left(1 + \frac{1}{4} \sin^2 \frac{\alpha}{2} + \frac{9}{64} \sin^4 \frac{\alpha}{2} + \ldots\right)$$

For small values of α, the effect of the second and subsequent terms of the series is negligible in practice. Nevertheless, for larger values of α, the period of oscillation does depend to some extent on α.

The result of the dimensional analysis does not conflict with the algebraic result. The effect of α on t is seen to be *indirect*: α enters the expression only in powers of sin ($\alpha/2$), and the sine of any angle is dimensionless. The fact that the exponent d in eqn 6.23 is zero indicates only that α does not enter the sought-for relation in such a way that a change of the scale of measurement for angle would affect any of the numbers in the relation.

It has been rare for writers on dimensional analysis to regard angle as having anything but a dimensionless magnitude. In cases where angle has a direct effect on the phenomenon considered there is evident advantage in regarding the magnitude of angle as fundamental. Provided that an apparent independence of angle is clearly recognized as indicating only that angle has no *direct* effect (yet may enter the expression in trigonometrical ratios) there is no harm in regarding the magnitude of angle as an additional fundamental magnitude in every case. Dimensional analysis can then indicate whether the effect (if any) is direct or indirect, whereas considering the magnitude of angle as dimensionless yields no information at all.

This conclusion illustrates a general truth. The dimensional formula of any magnitude indicates how the expression for that magnitude would be affected by changes of *scale* for the fundamental magnitudes. For example, the fact that the dimensional formula for acceleration is $[LT^{-2}]$ indicates that if the scale of time interval were changed from one of seconds to one of minutes (i.e. if the fundamental unit of time interval were made 60 times larger) then the numerical part of the expression for the magnitude of a particular acceleration would be multiplied by 60^2 (2 m/s², for instance, would become 7200 m/min²). The same dimensional formula similarly indicates the effect of a change in the length scale.

If angle enters a dimensional formula, the corresponding magnitude would be expressed differently if the scale of angle were changed, for example, from one of radians to one of degrees. A particular value of modulus of rigidity might be expressed as $8 \cdot 1 \times 10^{10}$ N/(m² rad) or $1 \cdot 414 \times 10^9$ N/(m² degree)—or even as $5 \cdot 09 \times 10^{11}$ N/(m² rev). But the expression for the period of oscillation of a simple pendulum, although affected by the magnitude of the amplitude, is not affected by the way in which that magnitude is expressed. For example, a simple pendulum, one metre long, swinging to 45° on each side of the vertical, has a period of 2·086 s in a locality where $g = 9\cdot81$ N/kg. If the same amplitude is expressed not on the degree scale but on the radian scale (as $\pi/4$ radians), the magnitude of the period is still expressed as 2·086 s. In this case, angle does not appear in the dimensional formula of the expression for the period of oscillation.

In a similar way, the magnitude of solid angle may be regarded as fundamental, instead of being considered simply as a ratio of a spherical surface area to the square of the corresponding radius. Alternatively, the magnitude of solid angle may be related to that of plane angle to give the dimensional formula [A²]. However, as the results seem to have no practical use, we shall not discuss further the dimensional formula of solid angle.

6.6 Dimensional Formulae for Thermal Quantities

As we have already seen (Section 6.2), the dimensional formula assigned to the magnitude of a physical quantity is ultimately a matter of convention. It depends on which magnitudes are selected as fundamental, and on how the derived magnitudes are defined in terms of the fundamental ones. Much discussion has taken place on these points in relation to thermal quantities. It is not necessary here to consider the many arguments brought forward, because opinion now has very largely settled in favour of regarding temperature as having a fundamental magnitude. The magnitudes of all thermal quantities may then be defined in terms of four fundamental magnitudes—those of length, mass, time interval and temperature. Quantity of heat is regarded as a form of energy, and so the dimensional formula of its magnitude is [ML²T⁻²].

However, it is worth mentioning that, by identifying absolute temperature with, for example, the mean kinetic energy of the molecules of the substance, it is possible to devise dimensional formulae for all thermal quantities in terms of the three 'mechanical' magnitudes [L], [M] and [T] only. Yet there is little, if any, practical advantage in doing this. Indeed, the 'Pi' theorem shows that, in dimensional analysis, the larger the number of fundamental magnitudes the more information is in general obtained.

The use of the four fundamental magnitudes, [L], [M], [T] and [θ] (this last symbol representing the magnitude of temperature), is now almost universal for thermal quantities, and has the advantage of being paralleled by the present definitions of *units* for these quantities. There is, it is true, no necessity for the fundamental magnitudes selected for dimensional formulae to be the same as those for defining units. In fact, anyone is entitled to choose and employ, in a particular context, what fundamental magnitudes he pleases, provided that the choice is consistent with definitions which agree with experimental observation. Yet the use of the same fundamental magnitudes for dimensional formulae as for units does aid the memory and it enables dimensional formulae to be readily recognized.

With [L], [M], [T] and [θ] as fundamental magnitudes, the dimensional formulae for various thermal quantities are obtained. For example:

Temperature	[θ]
Quantity of heat	[ML^{2}T^{-2}]
Enthalpy	[ML^{2}T^{-2}]
Specific enthalpy (i.e. enthalpy per unit mass)	[L^{2}T^{-2}]
Entropy	[ML^{2}T$^{-2}\theta^{-1}$]
Specific entropy (i.e. entropy per unit mass)	[L^{2}T$^{-2}\theta^{-1}$]
Gas constant R (defined by the equation $p = \rho RT$)	[L^{2}T$^{-2}\theta^{-1}$]
Specific heat capacity	[L^{2}T$^{-2}\theta^{-1}$]
'Latent heat' (of unit mass)	[L^{2}T^{-2}]

Thermal conductivity $= \dfrac{[\text{Heat}][\text{Length}]}{[\text{Cross-sectional area}][\text{Time interval}] \times [\text{Temperature difference}]}$

$\equiv$ [MLT$^{-3}\theta^{-1}$]

Thermal diffusivity
$= \dfrac{\text{Thermal conductivity}}{\text{Density} \times \text{Specific heat capacity}}$ [L^{2}T^{-1}]

Coefficient of expansion   [θ^{-1}]

As an example of dimensional analysis in which thermal quantities are involved, natural convection from a heated body will be considered. Convection may be defined as the transfer of heat by the movement of a fluid. When a body at a particular temperature is placed in a fluid at a different temperature, the density of the fluid near the body is altered slightly, and the fluid then moves under the influence of unbalanced buoyancy forces. If the movement of fluid relative to the body is due solely to this effect, the convection is known as *natural* (or *free*) convection. If, on the other hand, the relative motion between the fluid and the body is chiefly maintained by some other means (such as a fan or a pump), the convection is *forced*.

Here we confine attention to natural convection and we shall use dimensional analysis to examine the way in which various quantities affect the heat transfer coefficient, that is the heat lost in unit time from unit area of a heated body for a unit difference of temperature between the body and the main bulk of the fluid. We assume that the body is immersed in a single, homogeneous fluid. Thus, if it is a liquid, it contains no gas bubbles. Further, we suppose that the shape and orientation of the body are already specified; we are therefore not concerned with the effect of the geometry of the system. We also assume that no other solid bodies are near, and thus the movement of the fluid is not restricted.

Another necessary condition is that the thermal conductivity of the body is very much greater than that of the fluid. That is, we assume that the body itself offers negligible resistance to the transfer of heat, and thus the thermal conductivity of the body need not be considered as a significant parameter.

§6.6]   DIMENSIONAL FORMULAE FOR THERMAL QUANTITIES   77

The problem is of particular interest because it shows how other arguments may be used to supplement plain dimensional analysis so as to give a much more precise result.

The relevant quantities affecting the phenomenon are nine in number:

Heat transfer coefficient	h
Difference between temperature of body and temperature of main bulk of fluid	θ
Size of body, represented by one characteristic length	l
Thermal conductivity of fluid	k
Specific heat capacity of fluid (at constant pressure)	c_p
Density of fluid	ρ
Weight per unit mass	g
Temperature coefficient of density change	β
Viscosity of fluid	μ

If the magnitudes of these nine quantities are expressed in terms of the four fundamental magnitudes, length, mass, time interval and temperature, then the 'Pi' theorem indicates that five independent dimensionless parameters may be expected. Before embarking on the dimensional analysis, however, it is worth considering whether the number of independent dimensionless parameters may be reduced. Is it, for example, possible to use more than four truly independent fundamental magnitudes?

In this problem we may in fact use two additional fundamental magnitudes. First we may notice that the conversion of energy from one form to another does not enter this problem. Thus the equivalence of heat and mechanical energy does not here arise. Consequently we may consider the magnitude of quantity of heat, H, as defined quite independently of the magnitude of mechanical energy (or of the magnitude of anything else), and the dimensional formula [H] as quite distinct from the dimensional formula $[ML^2T^{-2}]$.

Secondly, since any accelerations of the fluid will be exceedingly small, Newton's Second Law may—most unusually—be disregarded. As in the problem discussed in Section 6.4.3.1, therefore, there is no necessary connection between the magnitudes of force and mass, and so that of force F may be taken as an additional fundamental magnitude. When certain relations (such as the 'mechanical equivalent of heat' and Newton's Second Law) do not enter a problem we handicap ourselves unnecessarily if we bring in such relations via the definitions of magnitudes.

With six fundamental magnitudes—those of length L, mass M, time interval T, force F, temperature θ, and quantity of heat H—now at our command we may construct the dimensional formulae of the nine variables:

Heat transfer coefficient	h	$[HL^{-2}T^{-1}\theta^{-1}]$
Difference of temperature between body and main bulk of fluid	θ	$[\theta]$
Size of body	l	$[L]$
Thermal conductivity of fluid	k	$[HL^{-1}T^{-1}\theta^{-1}]$
Specific heat capacity of fluid	c_p	$[HM^{-1}\theta^{-1}]$

Density of fluid	ρ	$[ML^{-3}]$
Weight per unit mass	g	$[FM^{-1}]$
Temperature coefficient of density change	β	$[\theta^{-1}]$
Viscosity of fluid	μ	$[L^{-2}FT]$

Notice that the dimensional formula of specific heat capacity is expressed in terms of [H], not in terms of mechanical energy as the usual units J/(kg K) suggest. Also, the dimensional formulae of g and μ avoid the usual equivalence of [F] and $[MLT^{-2}]$.

To use Buckingham's method a recurring set of six variables must be chosen which among them involve the six fundamental magnitudes [L], [M], [T], [F], [θ], [H]. We wish our final result to give h as a function of the other magnitudes, and so we do not include h in the recurring set. A suitable set not including h is $\theta, l, k, c_p, \rho, g$. One dimensionless parameter therefore takes the form

$$\Pi_1 = \theta^a l^b k^d c_p^e \rho^f g^j h$$

Its dimensional formula is

$$\left[\theta^a L^b \left(\frac{H}{LT\theta}\right)^d \left(\frac{H}{M\theta}\right)^e \left(\frac{M}{L^3}\right)^f \left(\frac{F}{M}\right)^j \left(\frac{H}{L^2 T\theta}\right)\right] \equiv [1]$$

Dimensional homogeneity in respect to [F] requires $j = 0$

Dimensional homogeneity in respect to [T] requires $-d - 1 = 0$,
  i.e. $d = -1$

Dimensional homogeneity in respect to [H] requires $d + e + 1 = 0$
  whence $e = 0$

Dimensional homogeneity in respect to [M] requires $-e + f - j = 0$
  whence $f = 0$

Dimensional homogeneity in respect to [θ] requires $a - d - e - 1 = 0$
  whence $a = 0$

Dimensional homogeneity in respect to [L] requires $b - d - 3f - 2 = 0$
  whence $b = 1$

$$\therefore \quad \Pi_1 = \frac{hl}{k}$$

Similarly it may be deduced that

$$\Pi_2 = \theta\beta \text{ and } \Pi_3 = \frac{\mu k}{c_p l^3 \rho^2 g}$$

A still further increase of precision is possible because it may be argued that the magnitudes g and β should appear together as a product. A density change $\Delta\rho$ causes a small volume V of the fluid to experience an upward (buoyancy) force $g(\Delta\rho)V$. However, the change of density results entirely from a change of temperature $\Delta\theta$, and so $\Delta\rho/\rho = \beta(\Delta\theta)$. Thus the buoyancy force is given by $\rho g \beta(\Delta\theta) V$. Gravity affects the situation only in determining this buoyancy force. It is conceivable that the change of density might affect the other fluid properties such as viscosity and thermal conductivity. Although both these properties depend considerably on temperature, however, they are unaffected by density changes alone (unless these are extreme). So, as g and β are effective only in

§6.6]   DIMENSIONAL FORMULAE FOR THERMAL QUANTITIES   79

determining the buoyancy force $\rho g \beta (\Delta \theta) V$, they influence the situation only as the product $g\beta$. The parameters Π_2 and Π_3 may therefore be combined as a ratio Π_2/Π_3, and the final result of the analysis is

$$\Pi_1 = \phi(\Pi_2/\Pi_3)$$

i.e.

$$\frac{hl}{k} = \phi\left(\frac{c_p l^3 \rho^2 g \beta \theta}{\mu k}\right) \tag{6.24}$$

This result, which is applicable to all fluids, is well corroborated by experiment. For the special case of diatomic gases, the parameter $c_p \mu/k$ (the Prandtl number—see Section 7.4) is practically the same for all such gases at a particular temperature. The expression 6.24 may then be written alternatively

$$\frac{hl}{k} = \phi_1\left(\frac{c_p l^3 \rho^2 g \beta \theta}{\mu k} \Big/ \frac{c_p \mu}{k}\right) = \phi_1\left(\frac{l^3 \rho^2 g \beta \theta}{\mu^2}\right)$$

Also, in earthly laboratories g is constant, and for a thermally perfect gas (that is, one obeying the relation $p/\rho T = $ constant, where p and T, respectively, represent the absolute pressure and absolute temperature of the gas) β is constant for a particular temperature. Thus the relation reduces to

$$\frac{hl}{k} = \phi_2\left(\frac{l^3 \rho^2 \theta}{\mu^2}\right) \tag{6.25}$$

for diatomic gases at a given temperature. Equation 6.25 suggests that when hl/k is plotted against $l^3 \rho^2 \theta/\mu^2$ (for a given geometrical arrangement) all points lie on a single curve. This is demonstrated remarkably well for long circular cylinders with horizontal axes, ranging from large steam pipes to fine wires.

6.7 Dimensional Formulae for Electric and Magnetic Quantities

The dimensional formulae of the magnitudes of electric and magnetic quantities require a fourth fundamental magnitude in addition to those of length, mass and time interval. The choice of this additional fundamental magnitude has been the subject of much discussion over a long period, and a wide diversity of views has been put forward. The suggestions receiving most support have been the magnitudes of the magnetic permeability of free space μ_0, the permittivity of free space ε_0, electrical resistance R, and electric charge Q. Any such choice of a fourth fundamental magnitude would suit the purposes of dimensional analysis. Opinion, however, has now largely settled on the use of electric charge Q, and we shall here consider dimensional formulae expressed in terms of [L], [M], [T] and [Q]. One practical advantage of regarding charge as having a fundamental magnitude is that the dimensional formulae of other magnitudes then seldom involve fractional exponents, whereas most other proposed choices of fundamental magnitudes produce dimensional formulae in which many of the exponents are fractions.

6.7.1 Dimensional Formulae of Magnitudes of Electric Quantities

From the four fundamental magnitudes of length, mass, time interval and electric charge, and from the various defining equations for the magnitudes of electric quantities, dimensional formulae may be deduced as shown in Table 6.5.

80   EXPRESSIONS IN PHYSICAL ALGEBRA   [Ch. 6

TABLE 6.5

Quantity	Symbol	Defining equation	Dimensional formula
Electric current	I	$I = \partial Q/\partial t$	$[T^{-1}Q]$
Electric field strength	E	$E = \underset{Q \to 0}{\mathrm{Lim}} \left(\dfrac{\text{Force on charge } Q}{Q} \right)$	$[FQ^{-1}] = [MLT^{-2}Q^{-1}]$
Potential difference	V	$V = \underset{Q \to 0}{\mathrm{Lim}} \left(\dfrac{\text{Work done in moving charge } Q}{Q} \right)$	$[ML^2T^{-2}Q^{-1}]$
Resistance	R	$V = IR$	$[VI^{-1}] = [ML^2T^{-1}Q^{-2}]$
Capacitance	C	$C = Q/V$	$[M^{-1}L^{-2}T^2Q^2]$
Inductance	L	$V = -L\partial I/\partial t$	$[ML^2Q^{-2}]$
Mutual inductance	M	$V = -M\partial I/\partial t$	$[ML^2Q^{-2}]$
Permittivity	ε	$F = \dfrac{Q_1 Q_2}{4\pi\varepsilon r^2}$	$[F^{-1}L^{-2}Q^2] = [M^{-1}L^{-3}T^2Q^2]$
Electric flux density (Electric displacement)	D	$D = \varepsilon E$	$[L^{-2}Q]$
Electric flux	Ψ	$\Psi = \int D\, dA$	$[L^{-2}QL^2] = [Q]$
Specific conductance (conductivity)	σ	$R = \dfrac{l}{\sigma A}$	$[lA^{-1}R^{-1}] = [M^{-1}L^{-3}TQ^2]$

6.7.2 Dimensional Formulae of Magnitudes of Magnetic Quantities

Historically, the study of magnetism began with observations of the forces between ferro-magnetic materials: under certain conditions, bodies consisting of substances such as iron strongly attract or repel each other. The connection between magnetism and electricity was demonstrated by the classic experiment of the Danish physicist H. C. Oersted in 1819 in which a compass needle was deflected when an electric current passed near by. It is now recognized that all forms of magnetism are the result of electric currents; that is, of charges in motion.

Thus no new fundamental magnitude is required to express the dimensional formulae relating to magnetic quantities. To derive these dimensional formulae in terms of [L], [M], [T] and [Q] we do, however, need some expression relating the magnitude of a magnetic quantity to the magnitudes of electric quantities. From a number of possibilities we may select the empirical relation

$$\delta F \propto BI \sin \theta \, \delta l \qquad (6.26)$$

Here δF represents the force exerted on a short length δl of a straight conductor carrying an electric current I when the conductor is in a magnetic field. The angle θ is that between the element δl of the conductor and the direction of the field at that position, and B represents the *magnetic flux density*, also termed *magnetic induction*. If the direction of the vector quantity B were different from that of the earth's magnetic field, the magnitude of B could be compared with the magnitude of the flux density of the earth's field by observing the deflection of a compass needle. However, by setting the coefficient of proportionality in the expression 6.26 equal to unity, this relation may be regarded as the defining equation for the magnitude of B. Then

$$B = \frac{\delta F}{I \sin \theta \, \delta l}$$

and so

$$[B] \equiv \left[\frac{F}{IL}\right] \equiv \left[\frac{MLT^{-2}}{T^{-1}QL}\right] \equiv [MT^{-1}Q^{-1}]$$

Magnetic flux Φ (Greek capital 'phi') is defined by the equation

$$\Phi = \int B \, dA$$

where all the elements of cross-sectional area dA are perpendicular to B. The corresponding dimensional formula of Φ is therefore $[ML^2T^{-1}Q^{-1}]$.

If the magnetic field is itself produced by an electric current I in a conductor, then each element δl of the conductor contributes an amount δB to the magnetic flux density at a specified point. Experiment shows that

$$\delta B \propto \frac{I \, \delta l \sin \theta}{r^2}$$

where r represents the distance between the element and the point considered, and θ represents the angle between the conductor and the line joining the element δl to the specified point. Experiment also indicates that the coefficient of proportionality depends on the medium between the conductor and the point in question. We therefore write

$$\delta B = \frac{\mu}{4\pi} \frac{I\,\delta l \sin \theta}{r^2}$$

where μ is termed the magnetic permeability of the medium. If there is only a vacuum between the conductor and the point considered, μ gives way to μ_0, the 'permeability of free space'. The 4π is introduced to 'rationalize' the units (see Section 3.5.5). By making the mathematical substitution

$$\delta H = \frac{I\,\delta l \sin \theta}{4\pi r^2}$$

the magnitude of the *magnetic field strength H* is defined. Thus

$$[H] \equiv \left[\frac{IL}{L^2}\right] \equiv [L^{-1}T^{-1}Q]$$

(Unfortunately, H is sometimes termed *magnetizing force*. It is not a force, and the dimensional formula is not $[MLT^{-2}]$.)

The dimensional formulae of other magnetic magnitudes may readily be derived as shown in Table 6.6.

TABLE 6.6

Quantity	Symbol	Defining equation	Dimensional formula
Magnetic flux density (or magnetic inductance)	B	$\delta F = BI \sin \theta\, \delta l$	$[MT^{-1}Q^{-1}]$
Magnetic flux	Φ	$\Phi = \int B\,dA$	$[ML^2T^{-1}Q^{-1}]$
Magnetic field strength	H	$\delta H = \dfrac{I\,\delta l \sin \theta}{4\pi r^2}$	$[L^{-1}T^{-1}Q]$
Magnetic permeability	μ	$\mu = B/H$	$[MLQ^{-2}]$
Magnetic moment	M	$M = $ (Torque on magnet)$/H$	$[ML^3T^{-1}Q^{-1}]$
Magnetomotive force† (m.m.f.)	F or M	$F = \int H\,dl$	$[T^{-1}Q]$
Reluctance	$\mathscr{R}$	$\mathscr{R} = \dfrac{\text{m.m.f.}}{\Phi}$	$[M^{-1}L^{-2}Q^2]$

7.3 EXAMPLES OF DIMENSIONAL ANALYSIS WITH ELECTRIC AND MAGNETIC QUANTITIES

Problems concerned with electric and magnetic quantities are often amenable to the methods of ordinary mathematics, and so dimensional analysis has not found such wide application in this field of study as in some others. However, one instance in which dimensional analysis has yielded valuable results is in the study of the operation of ultra-high-frequency thermionic vacuum tubes. We shall briefly consider this application of the technique now.

† An unfortunate name. The quantity is not a force.

The theory applies to grid-controlled vacuum tubes such as triodes, tetrodes or pentodes. The following conditions are assumed to hold:

(a) The current emitted by thermionic surfaces is limited by space charge (that is, by the cloud of electrons already in the neighbourhood of the surfaces).

(b) The initial velocity of the electrons is negligible.

(c) The maximum velocity attained by the electrons is small compared with the velocity of light. Thus we need not consider relativistic effects. (This condition presupposes potentials less than about 100 kV.)

(d) The tube is small enough for the time of propagation of electromagnetic waves inside it to be negligible compared with the period of oscillation.

(e) The motion of the electrons is not affected by magnetic fields.

(f) The cathode is not saturated.

(g) All electrodes are perfect conductors and so over each surface the electric potential is constant.

Let us first consider the efficiency of such a valve oscillator. The losses in the tube arise from the conversion of the kinetic energy of the electrons into heat when they hit the surfaces of the electrodes. Consequently, the efficiency depends on the motion of the electrons, and the basic equation governing the motion of each electron is

$$m\frac{d^2x}{dt^2} = -e\frac{dV_l}{dx}$$

where m and e, respectively, represent the 'rest' mass and charge of the electron which is moving in the x direction, and V_l represents the 'local' potential, that is the potential at the position of the electron. For continuous oscillations, the acceleration term d^2x/dt^2 depends on the frequency f of operation, and the potential gradient dV_l/dx depends on the e.m.f. of the power supply V and on the distance d between cathode and anode. We may suppose therefore that, for a tube of given type and geometrical shape, the efficiency η is a function of e, m, f, V and d. If the magnitudes of these six quantities are expressed in terms of the four fundamental magnitudes [L], [M], [T] and [Q], two dimensionless Πs may be expected. Since η is dimensionless, this can itself be one of the Πs; the other Π is formed from the remaining five magnitudes and may easily be shown to be $m^{1/2}fd/e^{1/2}V^{1/2}$. Thus the efficiency is a function of $m^{1/2}fd/e^{1/2}V^{1/2}$, or, since m/e has the same value for all electrons in all vacuum tubes, the result may be simplified to

$$\eta = \phi(fd/V^{1/2}) \tag{6.27}$$

The efficiency here referred to is the efficiency of the vacuum tube alone. The efficiency of the entire circuit depends also on impedances in the rest of the circuit. These impedances may, however, be adjusted so that the overall efficiency of the oscillator is a maximum, and this maximum efficiency (η_{max}) is then simply a function of $fd/V^{1/2}$. The relation between η_{max} and $fd/V^{1/2}$ for a geometrically similar set of tubes may readily be determined experimentally, and the result may then be used to predict the maximum efficiency obtainable from any tube in the set when used under given conditions.

Dimensional analysis enables us to derive other useful relations. At any given instant, we may suppose that the current density J (that is, current per unit surface area) leaving the cathode depends on e, m, d and V. (The frequency f is not included here because we are concerned only with the *instantaneous* value of J.) The current density also depends, however, on the permittivity of the medium ε, and this quantity must be included in our list of relevant quantities even though the value ε_0 for a vacuum is a 'universal constant'. If we take [L], [M], [T] and [Q] as fundamental magnitudes, then ε is not dimensionless. The dimensional formula for permittivity may be deduced from the defining equation $F = Q_1 Q_2/4\pi\varepsilon r^2$ which expresses the magnitude of the force F between two charges Q_1, Q_2 separated by a distance r. From this we see that

$$[\varepsilon] \equiv [F^{-1}L^{-2}Q^2] \equiv [M^{-1}L^{-3}T^2Q^2]$$

With ε_0 included we then have

$$J = \phi(e, m, d, V, \varepsilon_0)$$

From six quantities and four fundamental magnitudes, two Πs can be expected, and it may readily be shown that the result can be expressed in the form

$$\frac{Jm^{1/2}d^3}{e^{3/2}V^{1/2}} = \phi\left(\frac{V\varepsilon_0 d}{e}\right) \tag{6.28}$$

For the particular case of plane, parallel electrodes an algebraic solution is possible, and this yields

$$J = \frac{4}{9}\varepsilon_0\left(\frac{2e}{m}\right)^{1/2} V^{3/2} d^{-2}$$

This result is essentially that obtained by C. D. Child in 1911 and independently by Irving Langmuir in 1913, and it is now known as the Child–Langmuir law. The current density J is seen to be directly proportional to the permittivity ε_0. It is inconceivable that, in the more general case, J is affected differently by the permittivity of the medium. So we may suppose that, for any geometrical arrangement of the electrodes, $\phi(V\varepsilon_0 d/e)$ in eqn 6.28 has the form $KdV\varepsilon_0/e$, although the numerical coefficient K may well vary from one arrangement to another. We thus obtain

$$J = K\varepsilon_0(e/m)^{1/2}V^{3/2}d^{-2} \tag{6.29}$$

as the most general form of the Child–Langmuir law.† From this comes the important conclusion that, for any geometrical arrangement, the current density is proportional to $V^{3/2}$, and hence the total current also varies as $V^{3/2}$.

We can now use eqns 6.27 and 6.29 to establish other relations which are of value in studying the operation of valve oscillators.

Experience tells us that, for a given type of tube, there is only one value of $fd/V^{1/2}$ which gives a particular value of efficiency. So for this type of tube *operating at a given efficiency* we may write

$$fd/V^{1/2} = \text{constant } K_1$$

From the generalized Child–Langmuir law (eqn 6.29) we may write

$$Jd^2/V^{3/2} = \text{constant } K_2$$

† The foregoing line of argument may seem overcautious. Some people would perhaps bypass eqn 6.28 and boldly jump from the algebraic solution direct to eqn 6.29. They would no doubt rely on a kind of intuition that a change in the geometrical arrangement of the electrodes would affect only the value of the coefficient K. But such daring can be justified only by arguments from dimensional formulae, and any rigorous reasoning in fact requires eqn 6.28 or its equivalent.

Dividing $K_1{}^2$ by K_2 (so as to eliminate d) we obtain

$$\frac{f^2 V^{1/2}}{J} = \frac{K_1{}^2}{K_2} = \text{constant} \tag{6.30}$$

This result shows that, to obtain a particular value of efficiency, the useful plate potential V decreases rapidly as the frequency rises. This reduction of V could be remedied only by an increase of J.

Eliminating V by calculating $K_1{}^3/K_2$ gives

$$\frac{f^3 d}{J} = \text{constant} \tag{6.31}$$

Thus, for a given type of cathode and a given current density and efficiency, d varies inversely as f^3. This result implies very small values of d for high frequencies, and shows why a particular type of tube can be used only within narrow limits of frequency.

If we further suppose complete geometric similarity between different tubes we can obtain an expression relating power with frequency:

$$\text{Power} = VI = VJA \propto VJd^2 = \text{constant} \frac{J^2}{f^4} J \frac{J^2}{f^6} = \text{constant} \frac{J^5}{f^{10}} \tag{6.32}$$

Here the geometric similarity allows us to set the electrode surface area A proportional to d^2, and then we substitute for V and d from eqns 6.30 and 6.31. Although in practice complete geometric similarity is not normally achieved, eqn 6.32 does indicate why the power obtainable drops very rapidly when the frequency rises.

Finally we may take an example from magneto-fluid dynamics. Let us seek the relation giving the magnitude of the force exerted on a body immersed in a homogeneous viscous fluid of constant density subjected to electric and magnetic fields. The force F may be supposed to depend on nine other physical quantities as listed in Table 6.7.

TABLE 6.7

Quantity	Symbol	Dimensional formula
Force exerted on body	F	$[MLT^{-2}]$
Characteristic length of body	l	$[L]$
Relative velocity between body and fluid	v	$[LT^{-1}]$
Density of fluid	ρ	$[ML^{-3}]$
Viscosity of fluid	η†	$[ML^{-1}T^{-1}]$
Electric field strength	E	$[MLT^{-2}Q^{-1}]$
Magnetic field strength	H	$[L^{-1}T^{-1}Q]$
Magnetic permeability of fluid	μ	$[MLQ^{-2}]$
Permittivity of fluid	ε	$[M^{-1}L^{-3}T^2Q^2]$
Specific conductance of fluid	σ	$[M^{-1}L^{-3}TQ^2]$

† Here we use the symbol η for viscosity to avoid confusion with magnetic permeability.

From these ten quantities and four fundamental magnitudes we may expect six dimensionless products. We select as a suitable recurring set of variables the four, l, v, ρ, μ, since among them they involve each of the fundamental magnitudes and do not by themselves yield a Π. The first Π therefore has the form $l^a v^b \rho^c \mu^d F$. Since the magnetic permeability μ alone involves [Q] this Π can be dimensionless only if $d = 0$. The condition on [M] then gives $c = -1$; that on [T] gives $b = -2$; and that on [L] gives $a = -2$. Hence

$$\Pi_1 = \frac{F}{\rho l^2 v^2}$$

Similarly it may be shown that

$$\Pi_2 = \frac{\eta}{lv\rho}$$

$$\Pi_3 = \frac{E}{v^2 \rho^{1/2} \mu^{1/2}}$$

$$\Pi_4 = \frac{H\mu^{1/2}}{v\rho^{1/2}}$$

$$\Pi_5 = \varepsilon v^2 \mu$$

$$\Pi_6 = \sigma l v \mu$$

The sought-for relation connecting the ten magnitudes can be expressed in the form $\phi(\Pi_1, \Pi_2, \ldots, \Pi_6) = 0$. Or any combination of these Πs may be used in place of any one of them provided that the total number of independent parameters remains six.

Our concern has been to obtain the functional relation rather than to consider its physical interpretation. Yet we may note in passing that the product Π_2 is the reciprocal of the well-known parameter Reynolds number which is proportional to the ratio of |inertia force| to |viscous force| at a particular point in the flow. The significance of Π_4 is that it is proportional to the square root of the ratio |magnetic force| to |inertia force| at a particular point. And since, for any medium in which ε and μ are constant, $\varepsilon\mu = 1/c^2$, where c represents the velocity of propagation of electromagnetic waves (such as light) in the medium, $\Pi_5 = v^2/c^2$, that is, the square of the Lorentz number (see Section 7.4).

Certain combinations of these Πs are commonly used in magneto-fluid dynamics. For example, $\Pi_2^{-1}\Pi_4^2\Pi_6 = \sigma l^2 H^2 \mu^2/\eta$ is proportional to the ratio of |magnetically induced stress| to |viscous shear stress| at a particular point. The square root of this ratio has been termed the Hartmann number (see Section 7.4).

6.8 Closing Remarks on Dimensional Analysis

Dimensional analysis is a study of the restrictions placed on the form of an algebraic function by the requirement of dimensional homogeneity. That is, the analysis indicates the possible ways of grouping the magnitudes of those individual physical quantities which the investigator has supposed relevant to the phenomenon he is studying. Dimensional analysis cannot tell him whether the quantities he lists are in fact relevant. The method can seldom even indicate

§6.8]   CLOSING REMARKS ON DIMENSIONAL ANALYSIS   87

whether any relevant quantities have been omitted from the list, and it certainly cannot denote what they are. Making the initial list of quantities is the responsibility of the investigator himself. Dimensional analysis is a means of processing information, not providing it.

Results obtained inevitably depend on which quantities were at the outset considered to affect directly the phenomenon under investigation. If the original list of quantities fails to include one which is in fact relevant, then the result of the analysis will be incorrect. Usually, only experiment can indicate that something is wrong with the result. For example, the analysis may suggest that a dimensionless group Π_1 is a function of only one other dimensionless group Π_2. Experimental results would then be used to calculate corresponding values of Π_1 and Π_2 and these values could be plotted on a graph. If the points so obtained did not fit a single curve but showed a considerable 'scatter', the conclusion could be drawn that Π_1 was *not* a function of Π_2—or at least not of Π_2 alone. A search for an error, either in the analysis or in the assumptions on which it had been based, would then be required. If a previously omitted quantity were discovered, another attempt at the analysis could be made.

On the other hand, a quantity not relevant to the phenomenon might have been included in the original list. This redundant quantity might be rejected by the analysis itself (as was the mass of the pendulum in Sub-section 6.4.2.1), but more probably the analysis would yield an extra dimensionless product, say Π_x. Again, experiment would be required to reach the final truth. A series of experiments could be made in which Π_x was changed in magnitude while all other independent Πs were unaltered. If the experimental results showed that Π_x had no effect on the phenomenon, then Π_x could be omitted from the 'function' formula.

In verifying the predictions of dimensional analysis from graphs there are some hazards to which it is well to draw attention here. On occasion, the plotted points may closely fit a smooth curve, and yet the apparent correlation between the parameters may not be genuine.

A spurious correlation may arise when the dimensionless parameter plotted as ordinate contains some variable which is also in the parameter plotted as abscissa. There is nothing wrong in the presence of these common variables, but care should be taken in interpreting the graph because the fact that the 'scatter' of points is small could be due simply to the presence of a variable common to both parameters. For example, the parameter a/b may be plotted against b/c (or c/b). If much of the variation of a/b and b/c resulted from a wide range of b, a graph closely fitting the plotted points might be obtained even if there were no correlation at all between a and c. Also, a plot of ab against a (or of, say, $a + b$ against a) could produce a good fit to a single curve, even when a and b were entirely unconnected with each other.

Of course, few people would plot graphs of parameters as obviously related as these examples. Yet graphs of this sort are sometimes plotted *in effect* because the definitions of the parameters concerned are overlooked. Indeed, there have even been published instances of one variable, or a small group of variables, being plotted against itself because it appeared in both parameters and all other magnitudes were maintained constant.

Quite often, a relationship between two dimensionless parameters, which together embrace five or six individual magnitudes, is extracted from a graph of experimental data in which only two of the magnitudes were subject to appreciable variation. Thus, in seeking the physical significance of well-fitting graphs, care must be taken that false inferences are not drawn from a correlation which may be spurious.

Unusual choices of the scales to which points are plotted can also produce misleading results. The scatter of points is often apparently reduced when logarithmic scales are used, for example. Indeed, it is possible, by using ingenious combinations of magnitudes and choices of scales, to produce plausible-looking curves from data selected at random!

Although results from dimensional analysis can never be any more correct than the original assumptions about which quantities are relevant, it is a valuable means of supplementing the usual mathematical techniques. By itself, it will not provide a complete solution to a problem, but the partial solution it yields will indicate that, whatever the precise form of an unknown relation connecting the magnitudes involved, certain features of it, imposed by the need for dimensional homogeneity, are inescapable. Even when a problem is being examined by normal algebraic methods rather than by dimensional analysis, the requirement of dimensional homogeneity provides a very useful means of detecting errors. However, the major use of dimensional analysis is in guiding the experimenter so that he may obtain the maximum of information from the minimum number of experiments.

7

Physical Similarity

7.1 Introduction

Like dimensional analysis, physical similarity is of wide relevance. Practical problems are seldom solved by theoretical analysis alone, and it is frequently necessary to turn to experimental results to complete the study. Even if a complete quantitative theory has been worked out, experiments are still necessary to verify it, because theories are invariably based on certain assumptions which may not be precisely satisfied in practice.

Much of this experimental work is, of course, done either on the apparatus for which the results are required, or on an exact duplicate of it. But a large part of the progress made in many fields of study has come from experiments conducted on scale models. No aircraft is now built before exhaustive tests have been carried out on small models in a wind-tunnel; the behaviour and power requirements of a ship are calculated in advance from results of tests in which a small model of the ship is towed through water. Flood control of rivers, harbour works, chemical process plant and similar large-scale projects are studied in detail with small models, and the performance of turbines, pumps, propellers and other machines is investigated with smaller, model, machines. There are clearly great economic advantages in testing and probably subsequently modifying small-scale equipment; not only is expense saved, but also time.

In a number of instances tests are conducted with one working material, and the results applied to situations in which a different material is used. For example, information may be required on the flow of gases in part of a gas turbine, but in the experiments water may be used as the working fluid because of the ease with which the flow pattern may be made visible in water.

In all these examples, results taken from tests performed under one set of conditions are applied to another set of conditions. This procedure is made possible and justifiable by the laws of similarity. By these laws, what is observed in one set of circumstances may be related to phenomena occurring in other sets of circumstances. Comparisons are frequently made between a *prototype*, that is, the full-size aircraft, ship, river, structure, machine or other device, and the *model* apparatus. The use of the same materials for both prototype and model is not necessary. Nor is the model necessarily smaller than the prototype. The flow of fluid through an injection nozzle or carburettor, for example, would be more easily studied by using a model much larger than the prototype. So would the flow of gas between small turbine blades. Indeed, model and prototype may even be of identical size, although the two may then differ in regard to other factors such as velocity, properties of a fluid, temperature differences and so on.

For any comparison between prototype and model to be valid, the sets of conditions associated with each must be *physically similar*. 'Physical similarity' is a general term covering a number of different kinds of similarity. We shall first define physical similarity as a general proposition, and then consider separately the various forms it may take.

Two systems are said to be physically similar in respect to certain specified physical quantities when the ratio of corresponding magnitudes of these quantities between the two systems is everywhere the same.

If the specified physical quantities are *lengths*, the similarity is called geometric similarity. This is probably the type of similarity most commonly encountered, and, from the days of Euclid, most readily understood.

7.2 Types of Physical Similarity

7.2.1 Geometric Similarity

Geometric similarity is similarity of shape. The characteristic property of geometrically similar systems, whether plane figures, solid bodies or patterns of movement, is that the ratio of any length in one system to the corresponding length in the other system is everywhere the same. This ratio is usually known as the *scale factor*.

Geometric similarity is perhaps the most obvious requirement in a model system designed to correspond to a given prototype system. Yet perfect geometric similarity is not always easy to attain. For example, if, in the systems being compared, a fluid is flowing past solid boundaries, then not only must the overall shape of the model be geometrically similar to that of the prototype, but the inevitable roughness of the surfaces should also be geometrically similar. For a small model the surface roughness might not be reduced according to the scale factor—unless the model surfaces can be made very much smoother than those of the prototype. Some of the smaller parts of a model apparatus might be too fragile or too difficult to make if they were reduced by the same scale factor as larger parts. And in the study of the movement of sediment in rivers, for example, a small model might require—according to the scale factor—the use of a powder of impossible fineness to represent sand.

If for any reason the scale factor is not the same throughout, a distorted model results. For example, a prototype and its model may have overall shapes which are geometrically similar, but surface finishes which are not. In the case of very large prototypes, such as rivers, the size of the model is probably limited by the available floor space in the laboratory; but if the scale factor used in reducing the horizontal lengths is also used for the vertical lengths, the result may be a stream so shallow that the flow is unduly affected by surface tension forces. Here, then, a distorted model may be unavoidable.

7.2.2 Kinematic Similarity

Kinematic similarity is similarity of motion. This implies similarity of lengths (i.e. geometric similarity) and, in addition, similarity of time intervals. Then, since corresponding lengths in the two systems are in a fixed ratio and corresponding time intervals are also in a fixed ratio, the velocities of corresponding objects must be in a fixed ratio of magnitude at corresponding times. Moreover, accelerations of corresponding objects must be similar. If the ratio of corresponding lengths is r_l and the ratio of corresponding time intervals is r_t, then the magnitudes of corresponding velocities are in the ratio r_l/r_t and the magnitudes of corresponding accelerations in the ratio r_l/r_t^2.

A well-known example of kinematic similarity is found in a planetarium. Here the heavens are reproduced in accordance with a certain length scale factor, and in copying the motions of the planets a fixed ratio of time intervals (and hence velocities and accelerations) is used.

7.2.3 Dynamic Similarity

Dynamic similarity is similarity of forces. If two systems are dynamically similar then the magnitudes of forces at similarly located points in each system are in a fixed ratio. Consequently the magnitude ratio of any two forces in one system must be the same as the magnitude ratio of the corresponding forces in the other system. The forces acting in a system may be due to many causes: gravitational attraction, differences of pressure, elasticity of a material, fluid viscosity, surface tension, electromagnetism and so on. For perfect dynamic similarity, therefore, there are many requirements to be met, and it is usually impossible to satisfy all of them simultaneously. Fortunately, in many instances, some of the forces do not enter the problem being studied, or have only a negligible effect, and so it becomes possible to concentrate on the similarity of the most important forces.

A particular instance of dynamic similarity is *elastic similarity*, that is, similarity of the elastic stresses in solid materials. If the stresses at corresponding points in two solid structures (for example, a prototype bridge and its model) are to produce similar (small) strains, then the individual members of the structures must be made of such materials and have such cross-sections that two requirements are met: at corresponding points in the two structures the products of Young's modulus E and the section modulus Z must be in a fixed ratio, and the values of Poisson's ratio must be the same.

If the forces give rise to vibration of the structures, then complete physical similarity between the two systems requires, in addition, similarity of mass distribution.

7.2.4 Other Kinds of Similarity

In problems involving the transfer of heat, similarity between two systems usually requires *thermal similarity*; that is, differences of temperature between particular points in one system bear a fixed ratio to the differences of temperature between the corresponding points in the other system. Where chemical reactions are concerned, *chemical similarity* is usually required; that is, the concentration of a reactant at any point in one system is in a fixed ratio to the concentration of this reactant at the corresponding point in the other system.

Similarity of electric circuits requires, first, that particular points in one circuit are connected electrically in the same way as the corresponding points in the other circuit, and, second, that between the two circuits the magnitudes of all resistances must be in a fixed ratio, all inductances must be in a fixed ratio, and all capacitances must be in a fixed ratio.

For *electromagnetic similarity*, the electric field strength E and the magnetic field strength H must, at corresponding times, each be in a fixed ratio between corresponding points in the two systems being compared.

7.3 Mathematical Expression of Similarity Requirements

Let us suppose that the behaviour of two systems, such as a prototype and its model, is governed by an equation which may be put into the form

$$\phi(\Pi_1, \Pi_2, \Pi_3, \ldots) = 0$$

where $\Pi_1, \Pi_2, \Pi_3, \ldots$ represent dimensionless groups of variables. Any correct equation in which the magnitudes of all relevant quantities are explicitly included may be rearranged in this way. If the two systems are to be physically similar, then corresponding Πs must be the same in the two cases. The fact that the precise form of the function $\phi(\Pi_1, \Pi_2, \Pi_3, \ldots)$ may be unknown is of no importance: if each Π is the same for one system as for the other then the function must have the same value in each case and the equation is satisfied in each case. For physical similarity between two systems, then, the individual magnitudes for one system must bear such ratios to those for the other system that any Π containing the magnitudes has the same value in one system as in the other.

Tests conducted on the model system must show, for example, how the value of Π_1 depends on the values of Π_2, Π_3, Π_4, etc. If each of the independent parameters Π_2, Π_3, Π_4, etc. has the same value for the prototype as for the model, then the experimental results obtained with the model system are equally applicable to the prototype system.

Unfortunately, in practice there are often so many independent Πs influencing a phenomenon that it is impossible to satisfy all the requirements for physical similarity at the same time—unless the systems being compared are completely identical. Even so, it may be possible to achieve a useful degree of partial similarity by concentrating attention on those Πs which have most effect on the phenomenon being studied. Previous experience or some preliminary investigations may indicate that certain Πs have only a small effect: it may be feasible to forgo the matching of these Πs between the two systems in order to allow the equating of other, more important, Πs. In applying to the prototype system the results of experiments on the model system, it will then be necessary to make allowance in some way for the residual effect of those Πs which could not be equated between the two systems. This is usually known as allowing for *scale effect*: it often involves considerable experience in the particular field of study and knowledge of the physical processes affecting the phenomenon.

The application of physical similarity to particular situations is beyond the scope of this book. It is sufficient to note here that, ideally, comparison of two physical systems requires every relevant dimensionless parameter to have the same value in one system as in the other.

7.4 Named Dimensionless Parameters

In the investigation of physically similar systems, certain dimensionless parameters occur in a variety of problems. Many of these parameters arise so often that names have been given to them, and in this section we shall list and define a number of these named parameters.

Unfortunately, some of them have been given more than one name, and sometimes different writers have used the same name to refer to different parameters or to different forms of a parameter. Thus it is essential for readers to check the exact definition adopted by a writer in a given context.

In this list, what appears to be the most usual definition in English-speaking countries is given first; any less frequently used definitions then follow.

Very often, the name given to a parameter is that of the scientist who first made much use of it, or who was associated with the early development of that branch of the subject in which the parameter is chiefly used. The symbol then commonly adopted for the parameter consists of the first two letters of the name (except where this pair of letters is used for another parameter: thus the Graetz number has the symbol (Gz), instead of (Gr) which is used for the Grashof number). To indicate that the two letters are to be considered together as a single symbol and not as a product, they are often enclosed within parentheses; but the parentheses may be omitted when no confusion is likely.†

Sometimes names have been suggested for particular parameters, but the suggestions have not been adopted. These names which have not come into use are not included in this list.

Variation in the spelling of Russian names is often encountered: this results from the lack of a single standard convention for transliterating from the Russian (Cyrillic) alphabet. Some of these alternative spellings are noted here; other variants may be met but are not likely to cause confusion.

General Notation

(As far as is practicable the recommendations of the British Standards Institution are followed. This list refers only to the symbols used in the succeeding list of named dimensionless parameters: definitions here do not necessarily apply elsewhere in this book.)

A	area
a	velocity of sound in fluid
B	magnetic flux density
b	annular clearance
C_D	drag coefficient of body
c	velocity of light in a vacuum
c	specific heat capacity
c_b	specific vapour capacity (i.e. mass of vapour per unit mass of dry gas per unit pressure change)
c_p	specific heat capacity at constant pressure
c_v	specific heat capacity at constant density

† A variant of this notation, favoured particularly in America, uses the pair of letters as a suffix to N: e.g. N_{Gr} for Grashof number.

D	diameter of rotor or propeller
D_v	molecular diffusivity
d	diameter of pipe
d_p	diameter of particle
E	electric field strength
e	porosity (voidage)
F	force/length of bearing
f	friction factor, defined by $h_f = (4fl/d)(v^2/2g)$ (see Darcy coefficient in following list)
g	weight/mass
H	magnetic field strength
h	heat transfer coefficient, i.e. heat/(area × time interval × temperature difference)
h_f	head lost to friction
J	advance ratio of propeller
K	bulk modulus of fluid
k	thermal conductivity
k_m	mass transfer coefficient (mass flow rate per unit area per unit difference of concentration). Concentration is most often expressed as mass of solute/volume of solvent, and that definition is adopted here, thus giving k_m the dimensional formula [LT^{-1}]. Some writers, however, define concentration as mass of solute/mass of solvent; k_m then has the dimensional formula [$ML^{-2}T^{-1}$], and the definitions of dimensionless parameters involving k_m differ from those given here by a factor corresponding to the density of the solvent.
L	thickness of layer
l	representative length
m	hydraulic mean depth = cross-sectional area/perimeter in contact with fluid
$\dot{m}$	mass flow rate
N	rotational speed (expressed as revolutions/time interval). (Those parameters which involve N are dimensionless only if angle is regarded as having a dimensionless magnitude.)
P	power
p	absolute pressure
p^*	piezometric pressure $= p + \rho g z$ (ρ = constant)
Q	volume flow rate
R	gas constant defined by $p = \rho RT$
Re	Reynolds number
S	surface area/volume (of solid body)
T	absolute (i.e. thermodynamic) temperature
t	time interval
t_r	relaxation time of visco-elastic fluid
U	velocity of moving surface
U_i	reaction frequency for given chemical species
u	peripheral velocity of rotor
V	volume

§7.4] NAMED DIMENSIONLESS PARAMETERS 95

v	(representative) velocity of fluid (or of solid body relative to undisturbed fluid)
$\bar{v}$	average velocity of fluid over cross-section of pipe or channel
v_A	velocity of Alfvén magnetic waves $= B/(\rho\mu_e)^{1/2}$
y	distance from boundary surface
z	height above datum level

Greek symbols

α (alpha)	thermal diffusivity $= k/\rho c_p$
β (beta)	coefficient of volume expansion at constant pressure
γ (gamma)	surface tension
Δ (capital delta)	difference of (e.g. $\Delta\theta =$ temperature difference)
ε (epsilon)	permittivity
ε_0	permittivity of free space
ε_D	eddy mass diffusivity
ε_M	eddy momentum diffusivity
ε_T	eddy thermal diffusivity
θ (theta)	temperature (measured on arbitrary scale)
λ (lambda)	specific enthalpy of evaporation (formerly termed 'latent heat' of vaporization (or condensation))
μ (mu)	absolute (dynamic) viscosity *(often η)*
μ_e	magnetic permeability
μ_p	coefficient of rigidity (apparent viscosity)
ν (nu)	kinematic viscosity $= \mu/\rho$
π (pi)	3·14159 ...
ρ (rho)	density
σ (sigma)	electrical conductivity
τ (tau)	shear stress
τ_y	yield stress in very slow shear
ω (omega)	angular frequency

Suffixes (additional to those appearing in preceding list of symbols)

a	ambient
f	of fluid
g	of gas
int	at interface
l	of liquid
o	initial value
r	by radiation
s	of solid particle
v	of vapour
w	at, or of, wall
∞	at large distance from body
Superscript bar	mean value of (e.g. $\bar{x} =$ mean value of x)

In the following list, the name of the parameter† is given first, followed by the usual or recommended symbol. In some cases, where no suitable symbol is already in use, the author offers suggestions which are enclosed in square brackets. An occasional duplication of symbol (such as El, which has been used both for Ellis number and for Elsasser number) may be allowed if the two parameters are unlikely to be found in the same context.

Next comes the algebraic definition of the parameter, and then definitions of any special symbols not dealt with in the preceding general list. (The 'rationalized' system of equations—see Section 3.5.5—is employed for all electric and magnetic magnitudes, and the use of S.I. units is assumed.) An indication of the significance of the parameter is then often given: this frequently takes the form of a ratio of like magnitudes to which the named parameter is proportional (but not necessarily equal). For example, Reynolds number is proportional to the ratio of the magnitude of the inertia force to that of the viscous force at any given point in the flow of a fluid.

Finally, if the parameter is named after a person, this person is identified. If, however, the same name is used for more than one parameter, the identification is given only at the first mention of the name. In a few cases, it has not proved possible to discover after whom the parameter was named—at least, not without an amount of research quite out of proportion to the interest or usefulness of the information.

(The indicated pronunciation of foreign names is inevitably rather imprecise: the versions given here should be regarded as providing a reasonable approximation when the letters in italics are pronounced in the simplest and most obvious English manner.)

Absorption number, Ab

$k_L(z/D\bar{v})^{1/2}$, where k_L = individual liquid coefficient of absorption, z = length (from inlet) of surface covered by liquid film, D = diffusion coefficient of gas in liquid, $\bar{v}$ = mean velocity of liquid film over wetted wall column.

Acceleration number, $[Ac]$

$K^3/\rho g^2 \mu^2$

Enters problems of rapidly accelerated fluid flow; depends only on g and physical properties of fluid.

Addison shape number, $[Ad]$

1000 × 'Dimensionless' specific speed, q.v. (with N expressed as revolutions/time interval)

Herbert Addison (1889–   ), English engineer.

Advance ratio (of propeller), J

v/ND, where v = forward speed.

Aeroelasticity parameter, $[Ae]$

$\rho v^2/E$, where E = Young's modulus of solid material.

Aerodynamic load on structure/Elastic load within structure.

† Some writers use the words 'modulus' or 'criterion' in place of 'number' in many instances.

⟶ See also General Notation, pages 93–95 ⟵

§7.4] NAMED DIMENSIONLESS PARAMETERS 97

Alfvén number, Al (pronounced $Arlf'$-ven)
$$\frac{v_A}{v} = \frac{B}{v(\rho\mu_e)^{1/2}};$$ also defined as reciprocal of this.
Hannes Olof Gösta Alfvén (1908–   ), Swedish physicist.

Alfvén-Mach number—*see* Magnetic Mach number.

Archimedes number, Ar (pronounced Ark-im-mee'-$deez$)
$d_p{}^3 g(\rho_s - \rho_f)\rho_f/\mu^2$
A modified form, $V_s g(\rho_s - \rho_f)\rho_f/\mu^2$, has also been used.
Inertia force × Gravity force/(Viscous force)^2
Archimedes of Syracuse (287–212 B.C.), Greek mathematician.

Arrhenius group (pronounced A-ray'-nee-us)
E/RT, where $E =$ activation energy per unit mass.
Activation energy/Potential energy of gas. Determines rates of chemical reaction.
Svante August Arrhenius (1859–1927), Swedish chemist and physicist.

Bagnold number, $[Ba]$
$3C_D \rho_f v^2 / 4 d_p \rho_s g$
Drag force/Gravity force (in the intermittent transport of solid particles by fluids)
Ralph Alger Bagnold (1896–   ), English engineer.

Bairstow number
v/a_w (now obsolete)
Leonard Bairstow (1880–1963), English aeronautical engineer.

Bansen number, Ba
$h_r A_w / \dot{m} c$
Heat transferred by radiation/Thermal capacity of fluid.

Batchelor number, $[Bt]$
$vl\sigma / \mathbf{c}^2 \varepsilon$
George Keith Batchelor (1920–   ), Australian mathematician.

Beránek number, $[Be]$
$v_t{}^3 \rho_f{}^2 / \mu g(\rho_s - \rho_f)$, where $v_t =$ terminal falling velocity of solid particle.
(Inertia force)^2/(Viscous force × Gravity force)
[A particular case of the Lyashchenko number.]
Jaroslav Beránek, Czech chemist.

Bingham number, Bm (also known as Plasticity number)
$\tau_y L / \mu_p v$, where $L =$ width of channel.
Yield stress/Viscous stress in a Bingham plastic.
Eugene Cook Bingham (1878–1945), U.S. chemist.

Biot number, Bi (pronounced Be'-oh)
hl/k_s
Internal thermal resistance of solid body/Surface resistance (in problems of unsteady heat conduction).
In French and German literature, the definition of Biot number has often

→ See also General Notation, pages 93–95 ←

been the same as for Nusselt number—which must be carefully distinguished from Bi as defined above.
Jean Baptiste Biot (1774–1862), French physicist.

Biot number for mass transfer, Bi_m
$k_m L/D_{\text{int}}$, where D_{int} = molecular diffusivity at interface.
Mass transfer rate at interface between fluid and solid/Mass transfer rate in interior of solid wall of thickness L.

Blake number, Bl
$v/\nu(1 - e)S$
Inertia force/Viscous force in flow through granular materials.
F. C. Blake.

Bodenstein number, $[Bd]$
vL/D_a, where L = axial length, D_a = effective axial diffusivity.
[A special case of Péclet number for mass transfer, describing diffusion in beds of granular material.]

Boltzmann number $\equiv$ Thring radiation group.

Bond number, Bo
$(\rho - \rho_f)d^2 g/\gamma$, where ρ = density of bubble or droplet, ρ_f = density of surrounding fluid, d = diameter of bubble or droplet.
Gravity force/Surface tension force.
The name has also been used for the more general parameter $\rho g l^2/\gamma$.
Wilfrid Noel Bond (1897–1937), English physicist.

Bouguer number (1) $[Bg]$ (pronounced *Boo-gair'*)
$3C\lambda_r/2\rho_s \bar{d}_p$, where C = mass of dust/bed volume, λ_r = mean length of radiation path, ρ_s = density of dust.
Radiant heat transfer to dust in gas streams.

Bouguer number (2)
kl, where k = coefficient of attenuation for medium = coefficient of absorption + coefficient of scattering.

Boussinesq number $[Bq]$
$v/(2gm)^{1/2}$, where m = hydraulic mean depth of open channel.
(Inertia force/Gravity force)$^{1/2}$. Concerned with wave motion in open channels; similar to Froude number.
Joseph Boussinesq (1842–1929), French mathematician.

Brinkman number, Br
$\mu v^2/k\Delta\theta$
Heat generated by viscous action/Heat transferred by conduction.
Henri Coenraad Brinkman, Dutch physicist.

Bulygin number, Bu (pronounced *Boo-lee'-ghin*)
$$\frac{\lambda c_b \Delta p}{c(\theta_a - \theta_o)},$$ where c = specific heat capacity of moist body which is being dried by heat, θ_o = initial temperature of body.

$\rightarrow$ See also General Notation, pages 93–95 $\leftarrow$

Heat needed to vaporize that part of liquid which is removed by seepage/Heat needed to raise temperature of moist body to boiling point of liquid.

Camp number, Ca
$(PV/\mu Q^2)^{1/2}$, where $P=$ power dissipated by viscous action in volume V of fluid.
Elapsed time × average rate of shear in fluid. Criterion for degree of flocculation of suspended particles.
Thomas Ringgold Camp (1895–), U.S. engineer.

Capacity coefficient—*see* Flow coefficient.

Capillarity number
$\mu^2 K/\rho\gamma^2$
Depends only on physical properties of fluid; concerns action of surface tension in flowing fluid.

Capillarity-buoyancy number, M
$g\mu^4/\rho\gamma^3$. Reciprocal of this is also used.
Concerns effects of surface tension, viscosity and acceleration when liquid globules move through another fluid. μ, ρ refer to the 'outer' fluid, $\gamma =$ interfacial tension. *See also* Property group.

Capillary number
$\mu v/\gamma$
Viscous force/Surface tension force. Concerns atomization of liquids and two-phase flow through porous media.

Carnot number, Ca (pronounced *Kar'-no*)
$(T_2 - T_1)/T_2 =$ theoretical efficiency of Carnot cycle operating between two heat reservoirs at absolute temperatures T_1 and T_2.
Nicolas Léonard Sadi Carnot (1796–1832), French physicist.

Cauchy number, Ca (pronounced *Koh-she'*)
$\rho v^2/K$ [$=$ (Mach number)^2 if $K =$ *isentropic* bulk modulus].
Inertia force/Elastic force.
Augustin-Louis Cauchy (1789–1857), French mathematician.

Cavitation number, σ_c (also termed Leroux number)
$(p - p_v)/\tfrac{1}{2}\rho v^2$
The name is sometimes used for the Thoma number (q.v.).

Charge density number
$vl\rho_e/H$, where $\rho_e =$ excess charge density, i.e. charge/volume.

Clausius number, Cl (pronounced *Clow'-zi-ooss*)
$v^3 l\rho/k_t\Delta\theta$
Rudolf Julius Emanuel Clausius (1822–88), German physicist.

Colburn group or Colburn number $\equiv$ Schmidt number.

Colburn j factor (also termed J-factor for heat transfer)

→ See also General Notation, pages 93–95 ←

$$j = \frac{h}{\rho c_p v}\left(\frac{c_p \mu}{k_\mathrm{f}}\right)^{2/3} = (St)(Pr)^{2/3} = (Nu)(Re)^{-1}(Pr)^{-1/3}$$

Allan Philip Colburn (1904–55), U.S. chemical engineer.

Compressibility number = Harrison number.

Condensation number (1) (also termed Condensation film coefficient), Co

$$\frac{h}{k_\mathrm{f}}\left(\frac{v^2}{g}\right)^{1/3}$$

Condensation number (2) for vertical walls (also termed Vapour condensation group)

$l^3 \rho^2 g \lambda / k_\mathrm{f} \mu \Delta\theta$

Cowling number, Co

$B^2/\rho v^2 \mu_e = BH/\rho v^2$ [= (Alfvén number)^2]

The name has also been used for square root of this ratio and for $\sigma l B^2/\rho v$.

Thomas George Cowling (1906–), English mathematician.

Craya-Curtet number, C_t (pronounced Cry'-er-Coo'-er-tay)

$v_k/(v_d^2 - \tfrac{1}{2}v_k^2)^{1/2}$

where v_k = kinematic-mean velocity = $(\dot m_N + \dot m_I)/\rho A_T$

v_d = dynamic-mean velocity = $\{(i_N + \tfrac{1}{2}i_I)/\rho A_T\}^{1/2}$

$\dot m_N$ = mass flow rate through nozzle; $\dot m_I$ = mass flow rate of induced air stream

i_N = momentum flux through nozzle; i_I = momentum flux of induced air stream

ρ = gas density; A_T = cross-sectional area of furnace

Refers to radiant heat transfer in furnaces.

Antoine Joseph Edouard Craya (1911–), French engineer, and Roger Michel Curtet (1923–), French engineer.

Crispation group

$\mu\alpha/\gamma^* L$, where γ^* = undisturbed surface tension.

Concerns cellular convection currents caused by gradients of surface tension.

Crocco number, Cr

$$v/v_\mathrm{max} = \left\{1 + \frac{2}{(\gamma-1)M^2}\right\}^{-1/2}, \text{ where } v_\mathrm{max} = \text{maximum possible velocity}$$

of a perfect gas expanding isentropically, $\gamma = c_p/c_v$, M = Mach number.

Luigi Mario Crocco (1909–), Italian–U.S. physicist and aeronautical engineer.

Damköhler group (or parameter) I, Da_I (pronounced Dam-ker'-ler)

$U_i l/v$

Rate of chemical reaction in flowing gaseous mixture/Total mass flow rate.

Gerhard Friedrich Damköhler (1908–), German physical chemist.

Damköhler group (or parameter) II, Da_II

$U_i l^2/D_v$

Rate of chemical reaction in flowing gaseous mixture/Rate of molecular diffusion.

→ See also General Notation, pages 93–95 ←

Damköhler group (or parameter) III, Da_{III}
$qU_i l/vc_p\theta$, where q = heat liberated per unit mass, θ = temperature above datum value.
Heat liberated/Bulk transport of heat.

Damköhler group (or parameter) IV, Da_{IV}
$\rho q U_i l^2/k_f\theta$, where q and θ are as in Da_{III}.
Heat liberated/Heat conducted away.

Damköhler group (or parameter) V, $Da_V \equiv$ Reynolds number.

Darcy coefficient, f
In Britain, $f = 2gmh_f/l\bar{v}^2$, where l = length of pipe of constant cross-section. In America $f = 8gmh_f/l\bar{v}^2$ is more usual. Also known as friction factor, Fanning friction factor and Fanning number.
Henri Darcy (1803–58), French engineer.

Darcy number, Da
vl/k, where k = permeability of granular material.

Dean number, Dn
$(Re)(r/R)^{1/2}$, where $Re = \bar{v}d/\nu$, $r = d/2$, R = radius of curvature of centre-line of channel.
Expresses effect of centrifugal force on flow in curved pipes (of diameter d) or channels (of width d).
William Reginald Dean (1896–   ), English mathematician.

Deborah number, De (pronounced $Deb'\text{-}or\text{-}a$)
t_r/t_o, where t_o = residence time of material in process.
Hebrew prophetess mentioned in Judges, Chapters 4, 5.

Generalized Deborah number
$(I_e - I_w)^{1/2} t_n$, where I_e = invariant of rate of strain tensor, I_w = invariant of vorticity tensor, t_n = natural time of visco-elastic material.

Derjaguin (or Deryagin) number, De (pronounced $Derry\text{-}a'\text{-}ghin$)
$L(\rho g/2\gamma)^{1/2}$
Thickness of coating (L)/Capillary length.
Boris Vladimirovich Derjaguin (1902–   ), Russian physical chemist.

'Dimensionless' specific speed, K_n

$$\frac{NP^{1/2}}{\rho^{1/2}(gH)^{5/4}} \text{ for turbines;} \qquad \frac{NQ^{1/2}}{(gH)^{3/4}} \text{ for rotodynamic pumps,}$$

where H = head difference across machine. (Some writers express N in terms of $radians$/time interval.)
[K_n actually has the dimensional formula of angle.]
Characterizes shape of machine.

Discharge number—*see* Flow coefficient.

→ See also General Notation, pages 93–95 ←

Drag coefficient, C_D
Drag force on body/$\frac{1}{2}\rho v^2 A$, where A (usually) = maximum cross-sectional area of body perpendicular to v.
A special form for small particles with a steady settling velocity v is $(\rho_s - \rho_f)l_s g / \rho_f v^2$.

Drew number, $[Dr]$
$$\frac{(M_A - M_B)x_A + M_B}{(x_A - x_{Aw})(M_B - M_A)} \ln \frac{M_v}{M_w}$$
where M_A, M_B = 'molecular weight' of components A, B; M_v, M_w = 'molecular weight' of mixture in vapour and at wall; x_A = mole fraction of A in diffusing stream; x_{Aw} = mole fraction of A at wall.

Dufour number, Du
$$\frac{R}{c_p} k \frac{(n_1/n)_0}{(n_1/n)(n_2/n)} = \frac{R}{c_p} k \frac{(p_1/p)_0}{(p_1/p)(p_2/p)}$$
where k = (coefficient of thermal diffusion)/(coefficient of molecular diffusion); n_1, n_2 = number of molecules of components 1, 2 in given volume of binary gas mixture; $n = n_1 + n_2$; p_1, p_2 = corresponding partial pressures; $p = p_1 + p_2$.
Heat transfer by diffusion/Heat transfer by convection under conditions where linear transfer rates by diffusion and conduction are equal.
Louis Dufour (1832–92), Swiss physicist.

Dulong number, $Du \equiv$ Eckert number
Pierre Louis Dulong (1785–1838), French chemist and physicist.

Eckert number, Ec (Ek and Er have also been used). Also known as Dulong number.
$v_\infty^2 / c_p \Delta\theta$, where $\Delta\theta$ = difference of temperature between moving gas and adiabatic wall.
[= 2/Recovery factor]
Name has also been used for $v^2/c_p T$ (i.e. Kinetic energy/ Thermal energy of perfect gas).
Ernst Rudolf Georg Eckert (1904–   ), Czech–U.S. engineer.

Einstein number $\equiv$ Lorentz number
Albert Einstein (1879–1955), German physicist.

Ekman number, $[Ek]$
$(\nu/2\omega l^2)^{1/2}$, where ω = angular velocity of fluid. (The '2' is sometimes omitted.)
(Viscous force/Coriolis force)$^{1/2}$
Vagn Walfrid Ekman (1874–1954), Swedish physicist.

Elasticity number (1)
$\mu t_r / \rho r^2$, where r = radius of pipe.
Elastic force/Inertia force in visco-elastic flow.

→ See also General Notation, pages 93–95 ←

§7.4] NAMED DIMENSIONLESS PARAMETERS

Elasticity number (2)
$\rho c_p / \beta K$
Depends on physical properties only; concerns effect of elasticity of fluid on flow processes.

Electric field parameter
E/vB

Electrical characteristic number
$\rho(d\varepsilon/d\theta)l^2(\Delta\theta)E^2/\mu^2$
[Term has also been used for the Kronig number.]

Electrical Nusselt number
vl/D^* where $D^* = \frac{1}{2}(D^+ + D^-)$ and $D^+, D^- =$ diffusion coefficients of ions.

Electrical Reynolds number
This term has been variously applied to (i) $\varepsilon v/\sigma l$; (ii) vl/D^* (see Electrical Nusselt number); (iii) $\rho v^2/E\rho_e l \propto \rho v^4 \varepsilon l^2/i_s^2$ [$\propto$ Inertia force/Field force]; (iv) $E\rho_e l^2/\mu v \propto i_s^2/\mu \varepsilon v^3 l$ [$\propto$ Field force/Viscous force], where $\rho_e =$ charge density per unit volume, $i_s =$ streaming current.

Electromagnetic field number
$HE/\rho v^3$

Electromagnetic loading parameter = Electric field parameter.

Ellis number, El
$\mu_0 v/\tau_{1/2} d$, where $\mu_0 =$ limiting value of viscosity as shear $\to 0$; $\tau_{1/2} =$ shear stress when $\mu = \mu_0/2$.
Samuel Benjamin Ellis (1904–), U.S. chemist.

Elsasser number, El
$(v\sigma\mu_e)^{-1}$ [= Reynolds number/Magnetic Reynolds number]
Walter Maurice Elsasser (1904–), U.S. physicist.

Eötvös number, [Eo] (pronounced *Ert'-vursh*)
$(\rho - \rho_f)d^2 g/\gamma \equiv$ Bond number, q.v.
Roland von Eötvös (1848–1919), Hungarian physicist.

Euler number, Eu (pronounced *Oy'-ler*)
$\Delta p^*/\rho \bar{v}^2 = 2fl/d$, where $l =$ length of pipe of constant cross-section.
The name has also been used for $\Delta p^*/\frac{1}{2}\rho \bar{v}^2 = 4fl/d$ and for $\Delta p^* d/l\rho \bar{v}^2 = 2f$.
A modified version, $Eu' = H_l \rho_l g/v_g^2 \rho_g$, is used for the flow of vapour across mass transfer trays, where $H_l =$ head of gas-free liquid which hydrostatically balances the column of two-phase mixture on tray.
Leonhard Euler (1707–83), Swiss mathematician.

Evaporation number (1)
v^2/λ

Evaporation number (2)
$c_p/\beta\lambda$

→ See also General Notation, pages 93–95 ←

Evaporation-elasticity number
$K/\lambda\rho$

Expansion number
$\dfrac{gd}{v^2}\left(\dfrac{\rho_l - \rho_g}{\rho_l}\right)$, where $d=$ diameter of gas bubble in liquid, $v=$ velocity of bubble.
Buoyancy force/Inertia force.

Fanning friction factor, f (Also termed Fanning number)
See Darcy coefficient.
John Thomas Fanning (1837–1911), U.S. engineer.

Fedorov (or Fedorof) number (1), Fe (pronounced *Fyuh'-der-rawff*)
$\left\{\dfrac{4gd_p{}^3(\rho_s - \rho_f)\rho_f}{3\mu^2}\right\}^{1/3}\left[=\left(\dfrac{4}{3}\times \text{Archimedes number}\right)^{1/3}\right]$
Igor Mikhailovich Fedorov (–1951), Russian.

Fedorov number (2)
$\delta\beta^*$, where $\delta=$ Soret thermogradient coefficient, $\beta^*=$ Dufour coefficient.

Fenske number
Number of stages in a separation process.
Merrell Robert Fenske (1904–   ), U.S. chemical engineer.

Fine-structure constant, α
$e^2/2hc\varepsilon_0 \simeq 0{\cdot}0073$ where $e=$ charge on electron, $h=$ Planck's constant.
Reciprocal form was formerly used.

Fliegner numbers (pronounced *Fleeg'-ner*)
'Stagnation' number, $F(M) = \rho\bar{v}(c_p T_0)^{1/2}/p_0$
'Static' number, $f(M) = \rho\bar{v}(c_p T_0)^{1/2}/p$
'Impulse' number, $I(M) = \rho\bar{v}(c_p T_0)^{1/2}/(p + \rho\bar{v}^2)$
where $p_0=$ stagnation pressure (absolute) of gas stream, $p=$ static pressure (absolute), $T_0=$ stagnation temperature.
Albert Friedrich Fliegner (1842–1928), Swiss engineer.

Flow coefficient, ϕ
Q/ND^3 (in turbomachines)
Also termed capacity coefficient or discharge number. In particular applications, modified versions may be used: these invariably take the form $Q/(\text{a particular cross-sectional area in rotor} \times \text{circumferential velocity of blade at specified radius})$.

Fluidization number
v/v_o, where $v_o=$ fluid velocity at start of fluidization.

Fourier criterion for heat diffusion, Fo' (pronounced *Foo-ryay'*)
$D_v t/2\pi l^2$, where $t=$ period of temperature oscillation. (The 2π is sometimes omitted.)
Jean Baptiste Joseph Fourier (1768–1830), French mathematician and physicist.

→ See also General Notation, pages 93–95 ←

Fourier flow number, Fo_f
vt/l^2

Fourier number, Fo
$\alpha t/l^2$
Elapsed time t/(Time required to bring solid to final temperature based on steady heat conduction with a temperature difference equal to initial temperature difference in the unsteady heat conduction). Thus Fo indicates extent to which heating or cooling effects have penetrated through the body.

Fourier number for mass transfer, Fo_m
$k_m t/l$

Frank-Kamenetskiĭ number, δ (pronounced *Frank-Kam-men-yet'-skee*)
$$\frac{Q}{k}\frac{E}{RT_a^2}l^2 k_0 \exp\left(-\frac{E}{RT_a}\right)$$
where Q = heat liberated per unit mass of reactants per unit volume, E = activation energy per unit mass, k = thermal conductivity of reacting mixture, k_0 = pre-exponential constant in Arrhenius equation.
David Al'bert Frank-Kamenetskiĭ (1910–   ), Russian engineer.

Frequency parameter
$\omega l/v$ [= 2π × Strouhal number]

Frössling number for heat transfer, Fs_h (pronounced *Frerss'-ling*)
$Nu/Re^{1/2}$ for laminar flow over a flat plate,

$\dfrac{Nu - 2}{Re^{1/2}Pr^{1/3}}$ for turbulent flow round sphere,

where Nu = Nusselt number, Pr = Prandtl number.
Karl Gustav Nils Frössling (1913–   ), Swedish engineer.

Frössling number for mass transfer, Fs_m
$\dfrac{Sh - 2}{Re^{1/2}Sc^{1/3}}$ for mass transfer from sphere,

where Sh = Sherwood number, Sc = Schmidt number.

Froude number, Fr (pronounced *Frood*)
$v/(gl)^{1/2}$
(Inertia force/Gravity force)$^{1/2}$
The name has also frequently been used for the square of this ratio.
William Froude (1810–79), English naval architect.

Froude number (rotating), [Fr_R]
DN^2/g
An adaptation of square of preceding ratio for use with stirred liquids.

Galileo number, Ga (pronounced *Gal-li-lay'-oh*)
$l^3 g/v^2$
Inertia force × Gravity force/(Viscous force)^2
Galileo Galilei (1564–1642), Italian mathematician and astronomer.

→ **See also General Notation, pages 93–95** ←

Gay-Lussac number, [Gc]
$1/\beta\Delta\theta$
Joseph Louis Gay-Lussac (1778–1850), French chemist and physicist.

Goertler parameter, [Gl] (pronounced *Gert'-ler*)
$(v\theta/v)(\theta/r)^{1/2}$, where θ = momentum thickness of boundary layer, r = radius of (longitudinal) curvature of boundary.
Henry Goertler (1909–), German mathematician.

Goucher number, Go
$r(\rho g/2\gamma)^{1/2}$, where r = radius of wall or wire on which fluid coating is placed.
(Gravity force/Surface tension force)$^{1/2}$
Frederick Shand Goucher (1888–), Canadian physicist.

Graetz number, Gz (pronounced *Grairtz*)
$\dot{m}c_p/k_f l$, where l = length of heat transfer path (e.g. diameter of pipe).
Thermal capacity of fluid/Heat transferred by conduction.
Leo Graetz (1856–1941), German physicist.

Grashof number, Gr
$l^3 g \Delta\rho/\rho v^2$ (a form of the Archimedes number) = $l^3 g \beta \Delta\theta/v^2$ when density change is due only to temperature change. Often used in the equivalent form for a perfect gas, $l^3 g \Delta T/T v^2$, where $\Delta\theta = \Delta T$ = difference of temperature across fluid layer giving rise to free convection.
Inertia force × Buoyancy force/(Viscous force)^2
Franz Grashof (1826–93), German engineer.

Grashof number for mass transfer, Gr_m (also known as Diffusional Grashof number)
$l^3 g \beta_c (C_i - C)/v^2$, where β_c = coefficient of volume expansion due to solution (volume per unit mass), C_i = concentration (mass/volume) of solute at interface, C = concentration elsewhere.

Gukhman number, Gu (pronounced *Guch-man*, with 'ch' as in 'loch')
$(T_0 - T_m)/T_0$, where T_0 = absolute temperature of hot gas stream, T_m = absolute temperature of wet surface. (Since T_0 varies across the flow, an average value is normally used.)
Criterion for convective heat transfer in evaporation at constant pressure.
Aleksandr Adol'fovich Gukhman, Russian chemical engineer.

Guldberg-Waage group, GW (pronounced *Gool'-bairg-Vor'-gher*)
$y_1{}^{n_1} y_2{}^{n_2} \ldots - (x_1{}^{m_1} x_2{}^{m_2} \ldots)/k$
where $x_1, x_2 \ldots$ = mole fractions of reactants,
$m_1, m_2 \ldots$ = stoichiometric coefficients of the individual reactants,
$y_1, y_2 \ldots$ = mole fractions of products of reaction,
$n_1, n_2 \ldots$ = stoichiometric coefficients of the individual reaction products,
k = dimensionless partial equilibrium constant.
A criterion for similarity of chemical reactions.
Cato Maximilian Guldberg (1836–1902), Norwegian chemist and mathematician, and Peter Waage (1833–1900), Norwegian chemist.

→ See also General Notation, pages 93–95 ←

§7.4]   NAMED DIMENSIONLESS PARAMETERS   107

Gümbel number, *Gü* (1) (pronounced *Ghewm'-bel*)
$Fb^2/2\mu U r^2$, where r = shaft radius.
Ludwig Karl Friedrich Gümbel (1874–1923), German naval engineer.

Gümbel number, *Gü* (2)
$\mu\omega D/F$, where ω = angular velocity of shaft, D = diameter of shaft.

Hadamard number, *Ha* (pronounced *Ad'-dam-mar*)
$$\frac{3\mu_b + 3\mu_f}{3\mu_b + 2\mu_f}$$
where μ_b = viscosity of fluid in bubble, μ_f = viscosity of ambient fluid.
Jacques Salomon Hadamard (1865–1963), French mathematician.

Hall coefficient
This term usually refers to $E_y/J_x B_z$ (where J = current density and the suffixes indicate mutually perpendicular directions) but this is not a dimensionless parameter. However, the term is occasionally used for $\omega\tau$, where $\omega = eB/m^*$ = cyclotron resonance frequency, i.e. the angular velocity of electrons in a semiconductor about the direction of a magnetic field of flux density B, e = charge on electron, m^* = 'conductivity mass' of electron, τ = average relaxation time of electrons.
Edwin Herbert Hall (1855–1938), U.S. physicist.

Harrison number, [*Hr*]
$6\mu UL/p_a h_0^2$, where L = length of bearing pad in direction of motion, p_a = ambient or supply pressure, h_0 = thickness of lubricant film at exit.
William John Harrison (1884–1969), English mathematician.

Hartmann number, *Ha*
$Bl(\sigma/\mu)^{1/2}$
(Magnetically induced stress/viscous shear stress)$^{1/2}$
Julius Frederik Georg Poul Hartmann (1881–1951), Danish engineer.

Hatta number, *Ha*, [*Ht*]
$\gamma/\tanh \gamma$, where $\gamma = l(k_n C_B^{n-1}/D_A)^{1/2}$, k_n = reaction rate constant for *n*th order chemical reaction, C_B = average molar concentration of component B, D_A = diffusion coefficient of component A through other components, l = length of diffusion path (or $(v^2/g)^{1/3}$ for processes in packed towers).
Shirōji Hatta (1895–   ), Japanese chemical engineer.

Head coefficient, ψ
gH/N^2D^2, where H = head difference across turbomachine.
Also used:
gH/u^2 and $2gH/u^2$, where u = circumferential speed of blades at specified radius.

Heat transfer factor—*see* Colburn *j* factor.

Heat transfer number
$q/v^3 l^2 \rho$, where q = heat transfer/time.

→ See also General Notation, pages 93–95 ←

Hedström number, He (pronounced *Heerd'-strerm*)
$\tau_y l^2 \rho / \mu_p^2$ [concerned with flow of Bingham plastics]
Bengt Olof Arvid Hedström (1926-   ), Swedish chemist.

Helmholtz resonator group, $[Hh]$
$(d^3/V)^{1/2}/M$, where M = Mach number.
Frequency of pulsating combustion × Residence time
Herman Ludwig Ferdinand von Helmholtz (1821–94), German physicist.

Hersey number $[Hs]$
$F/\mu U$
Load force/Viscous force.
Mayo Dyer Hersey (1886-   ), U.S. mechanical engineer.

Hess number, $[He]$
$kl^2 C_0^{n-1}/a_m$, where k = reaction rate constant, n = order of chemical reaction, C_0 = initial concentration (mass/volume) of solid material, a_m = coefficient of potential conductivity of mass transfer for gaseous products of thermal decomposition of solid.
(Probably) Germain Henri Ivanovich Hess (1802–50), Swiss–Russian chemist.

Hodgson number, $[Ho]$
$Vf\Delta p^*/\bar{p}\bar{Q}$, where V = volume of system, f = frequency of pulsations of gas flow, Δp^* = drop of piezometric pressure due to friction and losses past obstructions.
John Lawrence Hodgson (1881–1936), English engineer.

Homochronicity number
Nt, where t = time required for rotating mixer to produce given degree of dispersion.

Homochronous number
vt/l, where t = time for fluid to move through characteristic distance l.

Hooke number ≡ Cauchy number
Robert Hooke (1635–1703), English mathematician and philosopher.

Hydraulic resistance group, Γc
$\Delta p/\rho_l gL$, where Δp = pressure drop across liquid on distillation tray, L = depth of liquid layer on tray.

Ilyushin number (pronounced *Ill-lew'-shin*)
$4(Re)\tau_D/3\rho v^2$, where τ_D = maximum dynamic slip stress. (For flow of viscoplastic liquid in circular pipe.)
Aleksei Antonovich Ilyushin (1911-   ), Russian mechanical engineer.

Interaction parameter, N
$\sigma B^2 l / \rho v$

J-factor for heat transfer ≡ Colburn j-factor (q.v.).

⟶ See also General Notation, pages 93–95 ⟵

§7.4]   NAMED DIMENSIONLESS PARAMETERS   109

J-factor for mass transfer, j_M
$$j_M = (k_m/v)(v/D_v)^{2/3}$$
Jakob modulus, Ja (pronounced Yak'-awb)
$c_l\rho_l\Delta\theta/\lambda\rho_v$, where $\Delta\theta$ = excess temperature of hot surface above boiling point of liquid.
Maximum bubble radius/Thickness of film of superheated liquid on hot surface.
Max Jakob (1879–1955), German–U.S. physicist.

Joule number, [Jo] (pronounced $Jool$)
$2\rho c_p \Delta\theta / \mu_e H^2$
Joule heating energy/Magnetic field energy.
James Prescott Joule (1818–89), English physicist.

(von) Kármán number (1), [Ka]
$2f(Re)^2$ or $f^{1/2}(Re)$ or $2f^{1/2}(Re)$ [Friction in pipes]
Theodore von Kármán (1881–1963), Hungarian–U.S. engineer.

(von) Kármán number (2)
$(k/v)(\tau_w/\rho)^{1/2}$, where k = mean height of surface roughness.

(von) Kármán number (3) $\equiv$ (Alfvén number)$^{-1}$.

Kirpichev heat transfer number, Ki_h (pronounced $Keer$-pee-$cheff'$)
$$\frac{h(\theta_s - \theta_a)l}{k_s \Delta\theta},$$ where θ_s = temperature at surface of body, $\Delta\theta$ = drop of temperature over length l within body.
Heat flux across surface of body/Heat flux within body.
Mikhail Viktor Kirpichev (1879–1955), Russian engineer.

Kirpichev mass transfer number, Ki_m
$\dot{m}l/\lambda_m(\theta_0 - \theta_p)$, where λ_m = coefficient of mass conductivity, θ_0 = initial mass transfer potential, θ_p = equilibrium value of mass transfer potential.

Kirpitcheff number, [Kp]
$(Re)^{2/3}C_D^{1/3}$

Knudsen number, Kn (pronounced $Nood'$-$s'n$)
Mean free path of molecules in gas at low density $\div l$.
Martin Hans Christian Knudsen (1871–1949), Danish physicist.

Knudsen number for diffusion, [Kn_D]
$eD_{AB}/q_D D_{KA}$, where D_{AB} = binary bulk diffusion coefficient for system AB, q_D = diffusion tortuosity, D_{KA} = Knudsen diffusion coefficient = $\frac{4}{3}$(Knudsen flow permeability constant) $\times$ (Equilibrium mean molecular speed of species A).
Bulk diffusion/Knudsen diffusion in granular bed.

Kondrat'ev number, Kn (pronounced Kon-$drat'$-$tieff$)
$$\frac{1}{\alpha S^2}\left(\frac{1}{\theta_a - \bar{\theta}}\right)\frac{d\bar{\theta}}{dt} = \frac{h}{kS}\left(\frac{\theta_a - \theta_s}{\theta_a - \bar{\theta}}\right),$$ where θ_s = temperature at surface of body, $\bar{\theta}$ = mean temperature of body.
Gorgii Mikhail Kondrat'ev (1887–   ), Russian engineer.

$\longrightarrow$ See also General Notation, pages 93–95 $\longleftarrow$

Kossovich number, Ko (pronounced $Kuh\text{-}saw'\text{-}veech$)
$\lambda \Delta u/c_p \Delta \theta$, where Δu = change of moisture content (mass of moisture/mass of dry medium).
Heat required for evaporation/Heat used in raising temperature of body.

Kozeny function, k (pronounced $Kozz\text{-}ay'\text{-}nee$)
$\dfrac{\Delta p^*}{\mu l} \dfrac{e^3}{(1-e^2)} \dfrac{1}{\bar{v}S^2}$, where Δp^* = drop of piezometric pressure over thickness l of porous material, $\bar{v}$ = mean component of fluid velocity in direction of l.
Josef Alexander Kozeny (1889–), Austrian engineer.

Kronig number, Kr
$l^2 \beta \Delta \theta E_s^2 N \{\alpha + (2m_0^2/3kT_\infty)\}/Mv^2$, where $\Delta \theta = T_w - T_\infty$ (small compared with T_∞), E_s = electric field strength at heated surface, N = Avogadro's constant, α = mean value of coefficient of polarization for a gas molecule, m_0 = electric dipole moment of molecule, k = Boltzmann constant, M = 'molecular weight' of gas.
Inertia force × Electrostatic force/(Viscous force)^2.
Describes effect of electrostatic field on convective heat transfer from heated surface.
Ralph de Laer Kronig (1904–), German–Dutch physicist.

Kutateladze number (1), Ku (pronounced $Koo\text{-}tat'\text{-}el\text{-}ad\text{-}zee$)
$JEl/\rho vh$, where J = current density, h = specific enthalpy of moving gas in which electric arc occurs
Samson Semenovich Kutateladze, Russian engineer.

Kutateladze number (2)
$\lambda/c_p(\theta - \theta_w)$

Lagrange group or Lagrange number (1), $[Lg]$ (pronounced $La\text{-}grarnzh'$)
$P/\mu l^3 N^2$, where P = power supplied to agitator of characteristic size l.
Joseph Louis Lagrange (1736–1813), Italian–French mathematician.

Lagrange number (2)
(Rate of molecular mass transfer + rate of eddy mass transfer) ÷ Rate of molecular mass transfer.

Lagrange number (3)
$\Delta p^* r/\mu \bar{v}$, where r = radius of pipe.
Independent of Re for laminar flow, and constant for a given system.

Laplace number, La
$(\Delta p)L/\gamma$, where Δp = pressure drop across liquid on distillation tray, L = depth of liquid layer on tray.
Pierre Simon Laplace (1749–1827), French mathematician.

Larmor number, $[Lr]$
Larmor radius ÷ l
Sir Joseph Larmor (1857–1942), Irish mathematician.

⟶ See also General Notation, pages 93–95 ⟵

Laval number, $[Lv]$

$v \Big/ \left(\dfrac{2\gamma}{\gamma+1} RT\right)^{1/2}$, where $\gamma = c_p/c_v$.

Velocity of gas/Critical velocity of sound.
Carl Gustaf Patrik de Laval (1845–1913), Swedish engineer.

Lebedev number, Le (pronounced *Leb'-bed-eff*)

$eb(\theta_{a0} - \theta_0)/\rho_s c_b p_0$, where $b = \dfrac{\partial}{\partial \theta}\{\text{concentration (mass/volume) of vapour}\}$

at constant pressure.
Flow due to expansion/Total macroscopic flow of vapour from porous body.
Panteleĭmon Dmitrievich Lebedev, Russian engineer.

Leroux number $\equiv$ Cavitation number.

Leverett function, j

$(k/e)^{1/2} p_c/\gamma$, where $k =$ permeability of porous material, $p_c =$ capillary pressure (i.e. difference of pressure across interface between immiscible fluids).
Characteristic radius of interfacial curvature/Characteristic size of pores.
[Two-phase flow through porous medium]
Miles Corrington Leverett (1910–   ), U.S. chemical engineer.

Lewis number, Le

$\rho c_p D_v / k = D_v/\alpha$, where c_p and k represent values for mixture if no chemical reaction occurred.
This is now the preferred form, but the reciprocal was formerly used.
Sometimes known as Lewis–Semenov number.
Warren Kendall Lewis (1882–   ), U.S. chemical engineer.

Turbulent Lewis number, Le_T

$\rho c_p \varepsilon_D / k_T$, where $k_T =$ eddy thermal conductivity.

Load parameter of journal bearing

$\mu U r^2 / F b^2$. A variant of Sommerfeld number.

Lock number, $[Lk]$

$(dC_L/d\alpha)\rho c r^4/I$, where $C_L =$ lift coefficient and $\alpha =$ angle of attack for blades of helicopter rotor, $c =$ chord length of blade, $r =$ radius of rotor, $I =$ moment of inertia of blade about flapping hinge.
Christopher Noel Hunter Lock (1894–1949), English mathematician.

Lock blade inertia number = Lock number $\div$ 8.

Lorentz number, Lo

$v/\mathbf{c}$, where $v =$ velocity of body.
Hendrik Antoon Lorentz (1853–1928), Dutch mathematical physicist.

Luikov number, Lu (also spelt Lȳkov) (pronounced *Lew'-koff*)

$k_m l/\alpha$

Mass diffusivity/Thermal diffusivity.
Alekseĭ Vasil'evich Luikov (1910–   ), Russian engineer.

$\rightarrow$ See also General Notation, pages 93–95 $\leftarrow$

Lukomskii number ≡ Luikov number.

Lundquist number, Lu (pronounced *Lernd'-kvist*)
$\sigma Bl(\mu_e/\rho)^{1/2}$
Stig A. S. Lundquist (1922–), Swedish physicist.

Lyashchenko number (pronounced *Lass-shen'-koh*)
$v^3\rho_f^2/\mu g(\rho_s - \rho_f)$
(Inertia force)^2/(Viscous force × Gravity force)

Lykoudis number, $[Ly]$ (pronounced *Lik-koo'-this*)
(Hartmann number)^2/(Grashof number)$^{1/2}$
Paul S. Lykoudis (1926–), Greek–U.S. engineer.

Lȳkov number—*see* Luikov number.

M group—*see* Capillarity-buoyancy number.

McAdams group, $[Mc]$
$h^4 l \mu_l \Delta\theta / k_l^3 \rho_l^2 g \lambda$
Constant for a given orientation of surface on which liquid film condenses.
William Henry McAdams (1892–), U.S. chemical engineer.

Mach number, Ma (but M is usually used in algebraic work) (pronounced *Mark*)
v/a
Ernst Mach (1838–1916), Austrian physicist.

Magnetic force parameter
$B^2 \sigma l/\rho v$
Magnetic body force/Inertia force.
Also termed Stuart number.

Magnetic Grashof number
$4\pi\sigma\mu_e v(Gr)$, where Gr = Grashof number = $l^3 g \Delta\rho/\rho v^2$.

Magnetic Mach number
v/v_A. Also $v/(a^2 + v_A^2)^{1/2}$ and $v/(a + v_A)$.
Also termed Alfvén-Mach number.

Magnetic number
$B(\sigma l/\rho v)^{1/2}$ = (Magnetic force parameter)$^{1/2}$

Magnetic Oseen number
$\frac{1}{2}\{1 - (v_A/v)^2\}\sigma\mu_e l v$
Magnetic force/Inertia force.
Carl Wilhelm Oseen (1879–1944), Swedish physicist.

Magnetic Prandtl number
$\mu_e \sigma v$ [= Magnetic Reynolds number/Re]
Depends only on properties of fluid.
Ludwig Prandtl (1875–1953), German engineer.

Magnetic pressure number
$\mu_e H^2/\rho v^2$ [= (Alfvén number)^2 = Cowling number]
Magnetic pressure/(2 × Dynamic pressure).

→ See also General Notation, pages 93–95 ←

§7.4] NAMED DIMENSIONLESS PARAMETERS 113

Magnetic Reynolds number [Re_m]
$\sigma \mu_e l v$
Mass transport diffusivity/Magnetic diffusivity; Velocity of fluid/Velocity of magnetic field.
Osborne Reynolds (1842–1912), English engineer.

Maievskii number $\equiv$ Mach number.

Marangoni number, [Mr]
$$\frac{\Delta \gamma}{\Delta \theta} \cdot \frac{\Delta \theta}{\Delta L} \frac{L^2}{\mu \alpha}$$
Carlo Giuseppe Matteo Marangoni (1840–1925), Italian physicist.

Margoulis number (pronounced *Mar-goo'-lees*) $\equiv$ Stanton number
Wladimir Margoulis (1886–   ).

Mass transfer factor—*see J*-factor for mass transfer.

Merkel number, Me (pronounced *Mairk'-el*)
$k_m A / \dot{m}_g$, where $A = $ total surface area of water in contact with gas.
Mass of water transferred in cooling per unit humidity difference/Mass of dry gas.
Friedrich Merkel (1892–   ), German engineer.

Miniovich number, Mn (pronounced *Meen-yaw'-veech*)
Sr/e, where $r = $ radius of particles in packed bed.
Ya. M. Miniovich.

Naze number, Na
v_A/a
Mlle Jacqueline Naze (1935–   ), French mathematician.

Newton number, Ne
$F/\rho v^2 l^2$, where $F = $ hydrodynamic drag force on body.
Sir Isaac Newton (1642–1727), English mathematician.

Nusselt film thickness parameter
$L(g/\nu_l^2)^{1/3}$
A special case of (Galileo number)$^{1/3}$ for falling films.
Ernst Kraft Wilhelm Nusselt (1882–1957), German engineer.

Nusselt number, Nu
hl/k_f (N.B. Not to be confused with Biot number)
Actual heat transfer in forced convection/Heat transfer which would occur by conduction across stationary fluid layer of thickness l.

Nusselt number for mass transfer, $Nu_m \equiv $ Sherwood number.

Ocvirk number, [Oc]
$$\frac{F}{\mu U}\left(\frac{2b}{L}\right)^2, \text{ where } L = \text{ axial length of bearing.}$$
Load force on bearing/Viscous force.
Fredrick W. Ocvirk (1913–67), U.S. mechanical engineer.

→ See also General Notation, pages 93–95 ←

Ohnesorge number, Z (pronounced O'-ner-sor'-$gher$)
$\mu/(\rho l \gamma)^{1/2}$ [= (Suratman number)$^{-1/2}$]
Viscous force/(Inertia force $\times$ Surface tension force)$^{1/2}$
Wolfgang von Ohnesorge.

Péclet number, Pe (pronounced Pay'-$clay$)
$lv\rho c_p/k_f = lv/\alpha$ [= $(Re)(Pr)$]
Bulk transport of heat in forced convection/Heat transfer by conduction.
Jean Claude Eugène Péclet (1793–1857), French physicist.

Péclet number for mass transfer, Pe_m
lv/D_v
Bulk mass transport/Mass transport by diffusion.

Pipeline parameter
$av_0/2H$, where a = velocity of water-hammer wave, H = static head.
Maximum pressure rise due to water hammer/(2 $\times$ static pressure).

Plasticity number $\equiv$ Bingham number.

Poiseuille number, Ps (pronounced $Pwah$-zoy'-yuh)
$v\nu/(\rho_s - \rho_f)gd_p^2$
Viscous force/Gravity force.
Jean Léonard Marie Poiseuille (1799–1869), French physiologist.

Poisson's ratio, ν
Lateral strain/Longitudinal strain.
Siméon Denis Poisson (1781–1840), French mathematician.

Pomerantsev number (pronounced Pom-$meer$-$rant'$-$seff$) $\equiv$ Damköhler group IV
Alekseĭ Aleksandrovich Pomerantsev, Russian engineer.

Posnov number, Pn (pronounced Paw-$snawf'$)
$\delta\Delta\theta/\Delta n$, where δ = Soret thermogradient coefficient, n = specific moisture content, i.e. mass of moisture per unit mass of completely dry gas.

Power coefficient or Power number
$P/l^5\rho N^3$, where l usually = D.
Drag force on rotating impeller/Inertia force.

Prandtl dimensionless distance, y^+
$(y/\nu)(\tau_w/\rho)^{1/2}$
Ludwig Prandtl (1875–1953), German engineer.

Prandtl number, Pr (σ has commonly been used as a symbol)
$c_p\mu/k_f = \nu/\alpha$
Momentum diffusivity/Thermal diffusivity. Depends only on properties of fluid.

Prandtl number for mass transfer $\equiv$ Schmidt number.

Diffusion Prandtl number
ν/D_v

$\rightarrow$ See also General Notation, pages 93–95 $\leftarrow$

Eddy or Turbulent Prandtl number, Pr_T
$\varepsilon_M/\varepsilon_T$ (for heat transfer in turbulent flows).

Effective or Total Prandtl number
$$\frac{\varepsilon_M + \nu}{\varepsilon_T + \alpha}$$
Total momentum diffusivity/Total thermal diffusivity for heat transfer in combined laminar and turbulent flows. (The Prandtl number $Pr = \nu/\alpha$ is sometimes termed the molecular Prandtl number to distinguish it from the eddy and effective Prandtl numbers.)

Prandtl velocity ratio, u^+
$v/(\tau_w/\rho)^{1/2}$
(Inertia force/Shear force at boundary)$^{1/2}$.

Predvoditelev number, Pd (pronounced *Prehd-voh-dee'-tyeh-lehv*)
$\Gamma l^2/\alpha T_0$, where $\Gamma =$ maximum rate of change of ambient temperature, $T_0 =$ initial (or other suitable known) temperature.
Rate of change of ambient temperature/Rate of change of temperature of body.
Aleksandr Savvich Predvoditelev (1891–), Russian physicist.

Pressure coefficient
$\Delta p/\rho v^2$
A special case of Newton number.

Pressure number
$p/\{g\gamma(\rho_l - \rho_g)\}^{1/2}$
Absolute pressure in system/Pressure jump at liquid surface.

Property group
$g\mu^4(\rho_f - \rho_s)/\rho_f^2 \gamma^3$, where $\rho_s =$ density of solid particle or of bubble.
A modification of capillarity-buoyancy number often giving better correlations with experiment. The reciprocal form has also been used.

Psychrometric ratio (for wet- and dry-bulb thermometry)
$h_c/k_m s$, where $h_c =$ heat transfer coefficient for convection, $s =$ humid heat, i.e. the heat required to give a unit rise of temperature to unit mass of dry air plus such water vapour as it contains.

Radiation number
$kK/\gamma\sigma T^3$, where $\sigma =$ Stefan–Boltzmann constant.

Radiation parameter, Φ
$\zeta\sigma T_w^3 m/k_f$, where $\zeta =$ coefficient expressing mean emissivity of internal surface of walls of channel, $\sigma =$ Stefan–Boltzmann constant.
Expresses influence of radiation on convective heat transfer in channel. A variant of Stefan number.

Ramzin number, $[Rz]$
Bulygin number/Kossovich number.

→ See also General Notation, pages 93–95 ←

Rayleigh number, Ra
$l^3\rho^2g\beta c_p\Delta\theta/\mu k_f$ [$= (Gr)(Pr)$]
This form is by far the most common, but the name has also been used for $qd^5\rho^2g\beta c_p/\mu k_f^2 x$, where q = rate of heat flow per unit area, x = distance from inlet of vertical tube. $(v^2\rho l/\gamma)^{1/2}$ [= Weber number] has occasionally been termed Rayleigh's parameter.
John William Strutt (1842–1919), the third Baron Rayleigh, English physicist.

Reaction enthalpy number
$(\Delta h)_A \Delta n_A/c_p\Delta T$, where $(\Delta h)_A$ = enthalpy of reaction/mass of A produced, n_A = mass fraction of A.
Change in reaction energy/Change in thermal energy.

Reactor dispersion group ≡ Reciprocal of Péclet number for mass transfer.

Recovery factor, [(RF)]
$2c_p\Delta\theta/v^2$ [= 2/Eckert number], where $\Delta\theta$ = difference of temperature between moving gas and adiabatic wall.
Actual temperature recovery/Theoretical temperature recovery for perfect gas.

Reech number
v^2/gl or gl/v^2.
Now superseded by Froude number.
Ferdinand Reech (1805–80), Alsatian engineer.

Reynolds number, Re
$vl\rho/\mu$
Inertia force/Viscous force.
Osborne Reynolds (1842–1912), English engineer.

Generalized Reynolds number for non-Newtonian fluids
$8\rho\bar{v}^2/\tau_w$
Applies to flow of non-Newtonian fluids in tubes of circular cross-section.

Reynolds number (rotating), [Re_R]
$\rho ND^2/\mu$

Richardson number, Ri
$-\dfrac{g}{\rho}\left(\dfrac{d\rho}{dz}\right) \Big/ \left(\dfrac{dv}{dz}\right)_w^2$, where z = height of layer (measured vertically upwards) in stratified flow.
Gravity force/Inertia force.
Lewis Fry Richardson (1881–1953), English physicist.

Romankov number, Ro
$(T_o - T_{pr})/T_o$, where T_o = absolute temperature of hot gas stream used in a drying process, T_{pr} = absolute temperature of product being dried.
Modified Romankov number, $Ro' = T_o/T_{pr}$
P. G. Romankov, Russian chemical engineer.

→ See also General Notation, pages 93–95 ←

Rossby number (1), *Ro*
$v/\omega l$, where ω = angular velocity of fluid.
Carl-Gustaf Arvid Rossby (1898–1957), Swedish–U.S. meteorologist.

Rossby number (2), *Ro*
$v/2\omega l \sin \alpha$, where ω = angular velocity of earth's rotation (or of other coordinate frame), α = angle between direction of (laminar) fluid flow and earth's axis of rotation.
Inertia force/Coriolis force.

Roughness factor or Roughness ratio
Average height of surface roughness ÷ d (or equivalent diameter for non-circular pipe).

Sarrau number
Name formerly used in France for Mach number.
Jacques Rose Ferdinand Émile Sarrau (1837–1904), French engineer.

Schiller number
$(Re/C_D)^{1/3}$.

Schmidt number (1), *Sc*
ν/D_v
Momentum diffusivity/Molecular diffusivity.
Ernst Heinrich Wilhelm Schmidt (1892–   ), German engineer.

Schmidt number (2)
$\nu \varepsilon / \sigma l^2$
Diffusivity of vorticity/Mass diffusivity of ions.

Eddy or Turbulent Schmidt number, Sc_T
$\varepsilon_M / \varepsilon_D$ (for mass transfer in turbulent flows).

Effective or Total Schmidt number
$$\frac{\varepsilon_M + \nu}{\varepsilon_D + D_v}$$
Total momentum diffusivity/Total mass diffusivity for mass transfer in combined laminar and turbulent flows.
(The Schmidt number $Sc = \nu/D_v$ is sometimes termed the molecular Schmidt number to distinguish it from the eddy and effective Schmidt numbers.)

Semenov number ≡ Lewis number.

Senftleben number, *Se* (pronounced *Senft-lay'-ben*)
$NE_s^2\{\alpha + (2m_0^2/3kT_\infty)\}/lMg$, where symbols have same meanings as for Kronig number.
Electrostatic force/Buoyancy force.
Hermann Max Senftleben (1890–   ), German physicist.

Shearer number
$M^{-1/3}$, where M = Capillarity-buoyancy number (q.v.).

→ See also General Notation, pages 93–95 ←

Sherwood number, Sh
$k_m l/D_v$
Mass diffusivity/Molecular diffusivity.
   Thomas Kilgore Sherwood (1903–   ), U.S. chemical engineer.

Size number (of turbomachine)
$D(gH)^{1/4}/Q^{1/2}$, where H = difference of head across turbomachine.
Also termed Specific diameter.

Smoluchowski number (pronounced *Shmol-oo-kof'-skee*)
(Knudsen number)$^{-1}$
   Maryan Ritter von Smolen Smoluchowski (1872–1917), Austrian physicist.

Sommerfeld fine-structure constant = Fine-structure constant, q.v.

Sommerfeld number, $[Sm]$ (The symbol Δ has been commonly used.)
$$\frac{\mu N D^2}{\bar{p} b^2} \text{ or } \frac{\mu N r^2}{\bar{p} b^2} \text{ or } \frac{Fb^2}{\mu U r^2} \text{ or } \frac{Fb^2}{2\mu U r^2}$$
where D = shaft diameter, $\bar{p}$ = mean pressure on bearing = F/D, r = shaft radius.
(The first of these ratios is the one mostly favoured in U.S.A.; the last is sometimes termed the Gümbel number.)
   Arnold Johannes Wilhelm Sommerfeld (1868–1951), German physicist.

Soret number, So (pronounced *So'-ray*)
$$\frac{k(n_2/n)_0}{(n_1/n)(n_2/n)},$$ where symbols have same meanings as for Dufour number.
   Charles Soret (1854–1904), Swiss physicist.

Spalding function, Sp
$-(\partial \theta/\partial u^+)_{u^+=0}$, where $\theta = (T - T_\infty)/(T_w - T_\infty)$, $u^+ = v/(\tau_w/\rho)^{1/2}$
Temperature gradient at wall, in dimensionless form.
   Dudley Brian Spalding (1923–   ), English chemical engineer.

Spalding number (1), Sp
$$\frac{hv}{vk\sqrt{(c_f/2)}} = \frac{hv}{k(\tau_w/\rho)^{1/2}},$$ where c_f = local skin-friction coefficient.

Spalding number (2), or Spalding transport number
$c_p \Delta T / \{\lambda - (q_r/\dot{m})\}$, where q_r = radiant heat flux, $\dot{m}$ = rate of mass transfer.
Change in thermal energy/'Latent heat' for evaporated material.

Specific diameter—*see* Size number.

Specific speed—*see* 'Dimensionless' specific speed.

Stanton number, St
$h/\rho c_p v$ [$= (Nu)/(Re)(Pr)$]
Quantity of heat actually transferred/Thermal capacity of fluid.
In France, the name has been used for (Prandtl number)$^{-1}$.
   Sir Thomas Edward Stanton (1865–1931), English physicist.

→ **See also General Notation, pages 93–95** ←

Stanton number for mass transfer, St_m
$k_m/v \; [= (Sh)/(Re)(Sc)]$

Stark number $\equiv$ Stefan number.

Stefan number, $[Sf]$
$\sigma T^3 l/k$, where $\sigma =$ Stefan–Boltzmann constant.
Rate of energy radiation/Rate of heat conduction.
Josef Stefan (1835–93), Austrian physicist.

Stewart number—*see* Stuart number.
(A mis-spelling apparently due to transliteration to, and then from, Russian.)

Stokes number (1), Sk
vt/l^2 where $t =$ time of vibration of particle in fluid, $l =$ characteristic length measurement of particle.
(Strouhal number)$^{-1}$ $(Re)^{-1}$
Sir George Gabriel Stokes (1819–1903), Irish mathematician and physicist.

Stokes number (2), Sk
$\omega l^2/v$, where $\omega =$ angular frequency of vibration of particle.
A variant of Stokes number (1).

Stokes number (3), Sk
$l \Delta p / \mu v$
Pressure force/Viscous force.

Strouhal number, Sr
fl/v, where $f =$ frequency of vibration.
The name has also been applied to the reciprocal of this ratio.
Vincenz Strouhal (1850–1922), Czech physicist.

Stuart number
(Hartmann number)$^2/(Re) = B^2 \sigma l / \rho v =$ Magnetic force parameter.
Magnetic body force/Inertia force.
John Trevor Stuart (1929–   ), English mathematician.

Suratman number, Su
$\rho l \gamma / \mu^2 \; [= $ (Ohnesorge number)$^{-2}]$
Inertia force $\times$ Surface tension force/(Viscous force)^2
P. C. Suratman.

Surface elasticity number

$$-\frac{\Gamma''}{D_s} L \frac{\partial \gamma}{\partial \Gamma''}$$

where $\Gamma'' =$ surface concentration of surfactant in undisturbed state, $D_s =$ surface diffusivity, $L =$ thickness of liquid layer.

Surface viscosity number
$\mu_s / \mu L$, where $\mu_s =$ 'surface viscosity', $L =$ thickness of liquid layer.

→ **See also General Notation, pages 93–95** ←

Taylor number (1), Ta
$\omega \bar{r}^{1/2} b^{3/2}/\nu$ or $\omega^2 \bar{r} b^3/\nu^2$ or $\omega r_i^{1/2} b^{3/2}/\nu$,
where ω = angular velocity of rotating cylinder of radius r_i, $\bar{r}$ = mean radius of annulus surrounding rotating cylinder.
Criterion for stability of Taylor vortices in annulus between two concentric cylinders when inner cylinder rotates.
Sir Geoffrey Ingram Taylor (1886–   ), English mathematician.

Taylor number (2)
$(2\omega L^2 \cos \theta/\nu)^2$ [= (Ekman number)$^{-4}$], where ω = angular velocity, θ = angle between axis of rotation and the vertical.
(Coriolis force/Viscous force)^2
Expresses effect of rotation on free convection.
Name has also been given to square root of this.

Taylor number (3) ≡ Sherwood number.

Temperature recovery factor—see Recovery factor.

Thiele modulus, m_T (pronounced *Tee'-ler*)
$\frac{1}{2} d_p (k/mD_v)^{1/2}$ = (Damköhler group II)$^{1/2}$,
where d_p = effective particle diameter of porous catalyst = $6/S$, k = reaction rate constant, m = average hydraulic mean depth of pores in particle, D_v = diffusion coefficient of reactants through fluid.
Ernest William Thiele (1895–   ), U.S. chemical engineer.

Thoma cavitation parameter or Thoma number, σ (pronounced *Toh'-mah*)
$(p_1 - p_v)/\Delta p$, where p_1 = absolute pressure at low-pressure side of machine, Δp = difference in pressure across machine.
Dietrich Thoma (1881–1943), German engineer.

Thompson number ≡ Marangoni number
A mis-spelling of Thomson.
James Thomson (1822–92), Irish engineer.

Thomson number (1)—*see above*.

Thomson number (2), Th
vt/l [Usually t = (vibration frequency)$^{-1}$. Then Th = (Strouhal number)$^{-1}$.]

Thring radiation group, $[Tg]$
$\rho c_p v/\varepsilon \sigma T^3$, where ε = emissivity of surface, σ = Stefan–Boltzmann constant.
Bulk transport of heat/Heat transferred by radiation.
Meredith Wooldridge Thring (1915–   ), English engineer.

Thring–Newby criterion
$\left(\dfrac{\dot{m}_1 - \dot{m}_0}{\dot{m}_0}\right) \dfrac{r}{l}$, where $\dot{m}_0$ = mass flow rate of fluid through nozzle, $\dot{m}_1$ = mass flow rate of surrounding fluid, r = nozzle radius, l = half-width of furnace.
Meredith Wooldridge Thring (1915–   ), English engineer, and
Maurice Purcell Newby, (1917–   ), English physicist.

→ See also General Notation, pages 93–95 ←

§7.4]   NAMED DIMENSIONLESS PARAMETERS   121

Thrust coefficient (of propeller)
$T_c = T/\rho v^2 D^2$, where T = thrust, v = forward speed;
or $C_T = T/\rho N^2 D^4$
$[C_T = T_c J^2]$.

Torque coefficient (of propeller)
$Q_c = Q/\rho v^2 D^3$, where Q = torque, v = forward speed;
or $C_Q = Q/\rho N^2 D^5$
$[C_Q = Q_c J^2]$.

Truncation number
$\mu w/p$, where w = rate of shear.
Shear stress/Normal stress.

Valensi number, $[Va]$
$\omega l^2/\nu$, where ω = angular oscillation frequency of object in fluid if μ were zero.
Jacques Valensi (1903–   ), French engineer.

Vapour condensation group—*see* Condensation number (2).

Velocity number $\equiv$ Magnetic Reynolds number.

von Kármán—*see* Kármán.

Weber number, We (pronounced *Vay'-ber*)
$v(\rho l/\gamma)^{1/2}$
(Inertia force/Surface tension force)$^{1/2}$
The name has also been given to the square of this ratio, and occasionally to the reciprocal. In Russia, it has been applied to $\gamma/\rho g l^2$.
Moritz Weber (1871–1951), German naval architect.

Weber number (rotating), $[We_R]$
$D^3 N^2 \rho'/\gamma$, where ρ' = effective density.
An adaptation of square of Weber number for use with stirred liquids.

Weissenberg number, $[Ws]$ (pronounced *Vice'-en-berg*)
$\omega_3 v/\omega_1 l$, where $\omega_3 = \int_0^\infty s G\, ds$, $\omega_1 = \int_0^\infty G\, ds$, G = relaxation modulus of linear visco-elasticity, s = recoverable elastic strain.
Visco-elastic force/Viscous force.
Karl Weissenberg (1893–   ), German-English physicist.

Generalized Weissenberg number
$I_e^{1/2} t_n$, where I_e = invariant of rate of strain tensor, t_n = natural time of visco-elastic material.

→ **See also General Notation, pages 93–95** ←

APPENDIX 1

Symbols, Dimensional Formulae and Units for Principal Physical Quantities

The defining equations given in the following table are intended as no more than brief reminders. They should not be taken as complete in themselves, since the symbols often require rigorous definition.

The dimensional formulae are based on fundamental magnitudes of length L, mass M, time interval T, plane angle A, temperature θ, electric charge Q and luminous intensity I. These fundamental magnitudes are not the only possibilities but are likely to be those most often called on in ordinary use. Dimensionless magnitudes are indicated by [1].

Quantity and recommended symbol for magnitude	Usual defining equation for magnitude	Dimensional formula	S.I. unit	Other units in common use†
Geometrical Quantities				
Length, l	—	[L]	metre, m	inch; foot; yard; mile; micron, $\mu = 10^{-6}$ m; ångström‡, $A = 10^{-10}$ m; light-year $= 9.4605 \times 10^{15}$ m; astronomical unit, a.u. $= 1.496 \times 10^{11}$ m; parsec, pc $= 3.084 \times 10^{16}$ m
Area, A	$A = l^2$	[L²]	m²	ft²; yard²; acre $= 4840$ yard²; barn, b $= 10^{-28}$ m²; are $= 100$ m²; hectare, ha $= 10^4$ m²
Volume, V	$V = l^3$	[L³]	m³	ft³; litre; gallon

† For conversion factors see Chapter 5.

‡ Properly pronounced *Orng'-strem*, but often anglicized to *ang'-strom*.

Quantity and recommended symbol for magnitude	Usual defining equation for magnitude	Dimensional formula	S.I. unit	Other units in common use†
Geometrical Quantities—*contd.*				
Second moment of area, Ak^2	$Ak^2 = \int r^2 \, dA$	$[L^4]$	m^4	$inch^4$; ft^4
Angle, θ (theta)	$\theta = $ Arc/Radius	$[1]$	radian, rad	degree
Solid angle, Ω (capital omega)	$\Omega = $ Area/Radius^2	$[A]$ $[1]$	steradian, sr	—
Kinematic Quantities				
Time interval, t	—	$[T]$	second, s,	minute, min; hour, h; day; week; year
Velocity, v	$v = dl/dt$	$[LT^{-1}]$	m/s	ft/s; km/h; mile/h
Acceleration, a	$a = dv/dt$	$[LT^{-2}]$	m/s^2	ft/s^2
Angular velocity, ω	$\omega = d\theta/dt$	$[T^{-1}]$ $[AT^{-1}]$	rad/s	degree/s; rev/s; rev/min
Quantities in Mechanics				
Mass, m	—	$[M]$	kilogram, kg	pound mass, lbm; slug; ton mass, tonm
Force, F	$F = ma$	$[MLT^{-2}]$	newton, N = kg m/s^2	poundal, pdl; pound-force, lbf; dyne; kilogram-force, kgf
Work or energy, E	$E = Fl$	$[ML^2T^{-2}]$	joule, J = m N	ft lbf; erg (= cm dyne); kW h = 3·6 MJ; electronvolt, eV
Power, P	$P = dE/dt$	$[ML^2T^{-3}]$	watt, W = J/s	ft lbf/s; horsepower, hp
Torque, T	$T = F \times $ Radius	$[ML^2T^{-2}]$	N m (NOT joule)	lbf ft

App. 1] SYMBOLS, DIMENSIONAL FORMULAE AND UNITS 125

Moment of Inertia I or Mk^2	$I = Mk^2 = \int r^2 \, dm$	$[ML^2]$	kg m²
Pressure, p	$p = F/A$	$[ML^{-1}T^{-2}]$	N/m²; lbm ft²; lbf/ft²‡; lbf/in²‡; atmosphere, atm; bar = 10^5 N/m²
Normal stress, σ (sigma) Shear stress, τ (tau)	$\left.\begin{array}{l}\sigma = F/A \\ \tau = F/A\end{array}\right\}$	$[ML^{-1}T^{-2}]$	N/m²; lbf/ft²; lbf/in²
(Dynamic) Viscosity, μ (mu)	$\mu = \tau/(\partial v/\partial y)$	$[ML^{-1}T^{-1}]$	N s/m² = kg/m s§; lbf s/ft²; poise, P = dyne s/cm² = gm/cm s
Density, ρ (rho)	$\rho = m/V$	$[ML^{-3}]$	kg/m³; lbm/ft³; g/cm³
Kinematic viscosity, v (nu)	$v = \mu/\rho$	$[L^2T^{-1}]$	m²/s; stokes, St = cm²/s; ft²/s
Surface tension, γ (gamma)	$\gamma = F/l$	$[MT^{-2}]$	N/m; dyne/cm

Quantities connected with Heat

Temperature (thermodynamic), T	—	$[\theta]$	kelvin, K (formerly °K); degree Rankine, °R
Quantity of heat, Q	Q = Energy	$[ML^2T^{-2}]$	joule, J; calorie; Btu; CHU; therm = 10^5 Btu
Specific heat capacity, c	$c = Q/m\Delta T$	$[L^2T^{-2}\theta^{-1}]$	J/kg K; cal/g degC; Btu/lbm degF
Entropy, S	$dS = dQ/T$	$[ML^2T^{-2}\theta^{-1}]$	J/K; Btu/°R
Specific entropy, s	$s = S/m$	$[L^2T^{-2}\theta^{-1}]$	J/kg K; Btu/lbm degR
Thermal conductivity, λ (lambda)	$\dfrac{Q}{t} = \dfrac{\lambda A(\Delta T)}{l}$	$[MLT^{-3}\theta^{-1}]$	W/m K; Btu/ft s degF

† For conversion factors see Chapter 5.
‡ The abbreviations 'psf' and 'psi', respectively, are also widely used, especially by Americans.
§ The special name 'poiseuille', with symbol Pl, has been proposed.

Electrical Quantities

Quantity and recommended symbol for magnitude	Usual defining equation for magnitude	Dimensional formula	S.I. unit	Other units in common use
Electric charge, Q	—	$[Q]$	coulomb, $C = A\ s$	faraday = 96500 C
Electric current, I	$I = \partial Q / \partial t$	$[QT^{-1}]$	ampere, A	
Electric potential, V	$V = \lim_{Q \to 0} \left(\dfrac{\text{Work}}{Q} \right)$	$[ML^2T^{-2}Q^{-1}]$	volt, $V = J/C$	
Electric resistance, R	$R = V/I$	$[ML^2T^{-1}Q^{-2}]$	ohm; $\Omega = V/A$	
Resistivity (specific resistance), ρ (rho)	$R = \rho l / A$	$[ML^3T^{-1}Q^{-2}]$	Ω m	Ω cm
Conductance, G	$G = R^{-1}$	$[M^{-1}L^{-2}TQ^2]$	reciprocal ohm, Ω^{-1} (sometimes termed 'mho'; the name siemens, S, is now recommended)	
Conductivity (specific conductance), σ (sigma)	$\sigma = \rho^{-1}$	$[M^{-1}L^{-3}TQ^2]$	$\Omega^{-1} m^{-1}$	
Current density, J or i	$J = I/A$	$[L^{-2}QT^{-1}]$	A/m^2	A/cm^2
Electric field strength, E	$E = V/l$	$[MLT^{-2}Q^{-1}]$	V/m	V/cm
Frequency, f	$f = \dfrac{\text{No. of cycles}}{t}$	$[T^{-1}]$	hertz, Hz = cycle/s	
Capacitance, C	$C = Q/V$	$[M^{-1}L^{-2}T^2Q^2]$	farad, $F = C/V$	
Permittivity, ε (epsilon)†	$F = Q_1 Q_2 / 4\pi \varepsilon r^2$	$[M^{-1}L^{-3}T^2Q^2]$	F/m	puff (a slang expression) = picofarad = 10^{-12}F
Electric flux density, D	$D = \varepsilon E$	$[L^{-2}Q]$	C/m^2	

† Data are normally given as *relative* permittivity, i.e. permittivity of medium/permittivity of free space.

Magnetic and Electromagnetic Quantities

[The sequence of defining equations here differs somewhat from that given in Section 6.7.2. The equations given here, being based on the concept of a turn of wire, are less general than those of Section 6.7.2, but are those on which the definitions of units have been founded.]

Number of turns, N	—	[1]	—
Magnetic flux, Φ (capital phi)	$V = -N\,\partial\Phi/\partial t$	$[ML^2T^{-1}Q^{-1}]$	weber, Wb = V s/turn	maxwell, Mx = 10^{-8} Wb (also known as a line of induction or, simply, line)
Magnetomotive force, F	$F = IN$	$[T^{-1}Q]$	ampere-turn, At	gilbert, Gb = $(10/4\pi)$At
Permeance, Λ (capital lambda)	$\Lambda = \Phi/F$	$[ML^2Q^{-2}]$	Wb/At	
Reluctance, $\mathcal{R}$ or S	$\mathcal{R} = F/\Phi$	$[M^{-1}L^{-2}Q^2]$	At/Wb	
Inductance, L	$V = -L\,\partial I/\partial t$	$[ML^2Q^{-2}]$	henry, H = V s/A	
Magnetic flux density, B	$B = \Phi/A$	$[MT^{-1}Q^{-1}]$	tesla, T = Wb/m^2	gauss†, G = 10^{-4}T (also termed lines/cm^2)
Magnetic field strength, H	$H = F/l$	$[L^{-1}T^{-1}Q]$	A/m = N/Wb	oersted‡, Oe = $(10^3/4\pi)$ A/m
Magnetic permeability, μ (mu)§	$\mu = B/H$	$[MLQ^{-2}]$	H/m	
Magnetic moment, M	$M = \text{Torque}/H$	$[ML^3T^{-1}Q^{-1}]$	Wb m	

Quantities connected with Illumination

Luminous intensity, I	—	[I]	candela, cd	candle-power ≙ candela
Luminous flux, F	$F = \int I\,d\Omega$	[I]	lumen, lm = cd sr	
Luminance, L	$L = I/A$	$[IL^{-2}]$	cd/m^2 (sometimes termed nit, nt)	stilb, sb = 1 cd/cm^2; apostilb, asb = $(1/\pi)$ cd/m^2; lambert, L = $(1/\pi)$ cd/cm^2; foot-lambert = $(1/\pi)$ cd/ft^2
Illumination, E	$E = dF/dA$	$[IL^{-2}]$	lux, lx = lm/m^2	foot-candle, fc = lm/ft^2; phot = lm/cm^2

† Pronounced *gowce*. The name has been applied to other units also—see Section 3.5.2.
‡ Pronounced *er'-sted*.
§ Data are normally given as *relative* permeability, i.e. permeability of medium/permeability of free space.

APPENDIX 2

Approximate Values of some Physical Constants and Common Properties

Weight per unit mass on earth's surface, g	9·81 N/kg (= 9·81 m/s²)
Standard value of weight per unit mass, g_0	9·806 65 N/kg (= 9·806 65 m/s² = 32·1740 ft/s²)
Velocity of light in a vacuum, c	3×10^8 m/s
Universal constant of gravitation, G	$6·67 \times 10^{-11}$ N m²/kg²
Universal (or molar) gas constant, R_0	8·314 kJ/kmol K
Standard temperature and pressure (s.t.p.)	273·15 K (= 0 °C) and $1·013 \times 10^5$ N/m²
Volume of 1 kmol of perfect gas at s.t.p.	22·41 m³
Electron-volt, eV	$1·602 \times 10^{-19}$ J
Charge on electron, e	$1·602 \times 10^{-19}$ C
Rest mass of electron, m_e	$9·109 \times 10^{-31}$ kg
Charge/mass of electron, e/m_e	$1·759 \times 10^{11}$ C/kg
Electron radius, r_e	$2·818 \times 10^{-15}$ m
Rest mass of proton, m_p	$1·673 \times 10^{-27}$ kg
Rest mass of neutron, m_n	$1·675 \times 10^{-27}$ kg
Magnetic permeability of free space, μ_0	$4\pi \times 10^{-7}$ H/m (exactly)
Permittivity of free space, ε_0	$8·854 \times 10^{-12}$ F/m
Avogadro constant, N	$6·023 \times 10^{26}$ kmol⁻¹
Boltzmann constant, k	$1·380 \times 10^{-23}$ J/K
Faraday constant, **F**	$9·65 \times 10^7$ C/kmol
Planck constant, **h**	$6·626 \times 10^{-34}$ J s
Rydberg constant, R_∞ ($= m_e e^4 / 8\varepsilon_0^2 \mathbf{h}^3 \mathbf{c}$)	$1·097 \times 10^7$ m⁻¹
Stefan–Boltzmann constant, σ	$5·67 \times 10^{-8}$ W m⁻² K⁻⁴
Bohr† magneton, β ($= e\mathbf{h}/4\pi m_e$)	$9·27 \times 10^{-24}$ m² A
Radius of first H orbit (Bohr radius)	$5·29 \times 10^{-11}$ m

Liquid Water

Density	10³ kg/m³
Viscosity at 20 °C	10⁻³ N s/m²
Specific heat capacity	4·187 kJ/kg K
Thermal conductivity at 15 °C	0·588 W/m K
'Latent heat' of boiling water at 100 °C	2·26 MJ/kg

Air

Density at s.t.p.	1·3 kg/m³
Viscosity at s.t.p.	$1·7 \times 10^{-5}$ N s/m²

† Pronounced *Bo'-er*.

Specific heat capacity at constant pressure	1 kJ/kg K
Specific heat capacity at constant volume	715 J/kg K
Ratio of principal specific heat capacities, γ	1·4
Thermal conductivity at s.t.p.	0·024 W/m K
Gas constant ($R = p/\rho T$)	287 J/kg K

APPENDIX 3

Some Other Units used in Scientific Work

Atmosphere (atm)	Unit of pressure now defined as exactly $1{\cdot}013\ 25 \times 10^5$ N/m². However, the name (with the abbreviation 'at') was formerly often used for 1 kgf/cm² = $9{\cdot}806\ 65 \times 10^4$ N/m².
Barye	A name sometimes used for dyne/cm² ($= 10^{-1}$ N/m²).
Bit	An abbreviation of 'binary digit'. It is the unit of information used in computers.
Carat	(1) A mass of approximately 200 mg, used in 'weighing' precious stones. The present internationally agreed value is exactly 200 mg. (2) A number indicating the purity of gold. Absolute purity is represented by 24 carat; 16 carat gold, for example, is an alloy of 16 parts (by mass) of gold and 8 parts of other material.
Curie (Ci)	Unit of radioactivity. Quantity of any radioactive nuclide in which the number of disintegrations is $3{\cdot}700 \times 10^{10}$ per second. The millicurie = 10^{-3} curie is more often used.
Cusec	Unit of volume rate of flow of a fluid = 1 ft³/s.
Darcy	Unit of permeability of a porous medium to fluid flow. The magnitude of permeability k is defined by the equation $k = \mu Q/A(dp^*/dl)$, where Q represents the rate of flow of fluid, of viscosity μ, through a block of material of cross-sectional area A when the gradient of piezometric pressure (in the general flow direction) is dp^*/dl. 1 darcy = 10^{-7} N/atm $\simeq 0{\cdot}987 \times 10^{-12}$ m².
Day, sidereal	Mean time taken for earth to complete one revolution. Equal to 86 194·0906 seconds (compare with solar day).
Day, solar	Mean time interval between successive appearances of sun in the same apparent position (e.g. due south). Equal to 86 400 seconds.
Decibel (dB)	Unit of power ratio used mainly in acoustics. The powers P_1 and P_2 of two sources differ by n decibels when $$n = 10\ \log_{10}(P_1/P_2).$$ (The smallest change of sound intensity detectable unaided by the normal human ear is about 1 dB.)
Dioptre	Unit for the reciprocal of the focal length of a lens or mirror. Equals (metre)⁻¹. Taken as positive for convex lenses or concave mirrors, and negative for concave lenses or convex mirrors.
f number of lens	Relative aperture expressed as focal length of lens/diameter of aperture. In photography, the f number is proportional to

	the square root of the exposure time (other things being unchanged).
Fathom	A unit of depth at sea. Equals 6 ft = 1·8288 m.
Fermi	Usually 10^{-15} m, but the name has sometimes been used as an alternative to 'barn' for an area of 10^{-28} m².
Galileo or gal	A unit of acceleration = 1 cm/s².
Grade	One-hundredth of a right angle.
Hectare	Unit of land area = 10^4 m².
Kilopond (kp)	Name used in Germany and Eastern Europe for kilogram-force (kgf).
Knot	Speed of 1 nautical mile (6080 ft = 1853·184 m) per hour. The international knot is 1852 m/h exactly.
Micron (μ)	Name formerly used for the micrometre (= 10^{-6} m).
Phon	A unit of loudness level. A sound has a loudness of n phons if it is judged by listeners (in standard listening conditions) to be equal in loudness to a simple tone, of frequency 1000 Hz, for which the sound pressure level is n decibels above 20 μN/m².
Rad	Unit of absorbed dose of an ionizing radiation. Defined as energy given to an element divided by the mass of the element. 1 rad = 10^{-2} J/kg.
Rayl	Unit of specific acoustic impedance. This quantity is defined as the complex representation of sound pressure (assumed to be varying sinusoidally) at a point in a sound wave divided by the complex representation of the velocity of particles at that point.
Reyn	A unit of viscosity equal to lbf s/in².
Rhe	A unit of fluidity (i.e. {viscosity}$^{-1}$) equal to cm²/s dyne. Sometimes termed 'reciprocal poise'.
Roentgen or Röntgen R (pronounced *rernt'-ghen*)	
	Unit of exposure to ionizing radiation. It is defined as the amount of photon radiation (i.e. primarily X-rays or gamma rays) which produces a certain degree of ionization in air. This amount of ionization was formerly specified as that produced when the total electric charge of the ions of one sign was 1 e.s.u. of charge in 1·293 mg of dry air (which is the mass of 1 cm³ of dry air at s.t.p.). The amount of ionization is now specified as 2·58 × 10^{-4} coulombs exactly per kilogram of dry air.
Torr	Unit of fluid pressure equal to 1/760 of a standard atmosphere. It is therefore very closely equal to 1 mm of mercury.

Further Reading

The relevant literature is so vast that anything like a complete list is quite impracticable. However, further discussion of some of the topics mentioned in Chapters 1–5 may be found in:

FEATHER, N. *An Introduction to the Physics of Mass, Length and Time*, Edinburgh University Press (1959) and Penguin, Harmondsworth (1961)
HVISTENDAHL, H. S. *Engineering Units and Physical Quantities*, Macmillan, London (1964)
JERRARD, H. G. and MCNEILL, D. B. *A Dictionary of Scientific Units*, 2nd edn, Chapman & Hall, London (1964)
Various authors. 'A Discussion on Units and Standards', *Proc. Roy. Soc.*, **A 186**, 149–217 (1946)

More advanced:
EISENBUD, L. 'On the Classical Laws of Motion', *Amer. J. Phys.*, **26**, 144–159 (1958)
ELLIS, B. D. *Basic Concepts of Measurement*, Cambridge University Press (1966)
JAMMER, M. *Concepts of Mass in Classical and Modern Physics*, Harvard University Press (1961) and *Concepts of Force*, Harvard University Press (1957)

The subjects Dimensional Analysis and Physical Similarity have a particularly extensive literature. C. Focken's *Dimensional Methods and their Applications*, Arnold, London (1953), covers most of the material published up to that date, and that book should be consulted for further references. P. W. Bridgman's *Dimensional Analysis*, 2nd edn, Yale University Press (1931), is still a classic of the subject, and he also contributed the article 'Dimensional Analysis' to the current (1970) edition of the *Encyclopaedia Britannica*. Among more recent works are B. D. Ellis's 'On the nature of dimensions', *Philosophy Sci.*, **31**, 357–380 (1964), W. J. Duncan's *Physical Similarity and Dimensional Analysis*, Arnold, London (1953), which is fairly advanced mathematically but has (to my mind) some questionable physical concepts, and J. Palacios' *Dimensional Analysis*, Macmillan, London (1964), which perhaps should be read in conjunction with criticisms of it in *Bull. Inst. Phys., Lond.*, **17**, 120–1 (1966).

S. J. Kline's *Similitude and Approximation Theory*, McGraw-Hill, New York (1965), and L. I. Sedov's *Similarity and Dimensional Methods in Mechanics*, Academic Press (Infosearch), London (1959), both contain useful material, although they should not be accepted uncritically.

List of Unit Symbols

Collected together here are the symbols for those units mentioned in this book. Combinations of prefixes and units are in general not included in this list, but exception is made for a few common examples such as cm and mm. Symbols for prefixes are listed separately at the end.

Those symbols preceded by an asterisk were formerly widely used, but are not now the standard symbols.

For information about the units, reference should be made to the general index.

Physical Quantities			
A	ampere	e.m.u.	electromagnetic unit
Å	ångström	e.s.u.	electrostatic unit
a	are; annum = year	eV	electron-volt
*amp	ampere		
amu	atomic mass unit	F	farad
asb	apostilb	°F	degree Fahrenheit
*AT, At	ampere turn	fc	foot-candle
atm	atmosphere	Fr	franklin
a.u.	astronomical unit	ft	foot
		ft-L	foot-lambert
b	bar; barn	G	gauss
Bi	biot	g	gram
Btu	British thermal unit	gal	gallon; galileo = cm/s^2
		*gall	gallon
C	coulomb	Gb	gilbert
°C	degree Celsius	gf	gram-force
*c	curie	*gm	gram
*c/s	cycle/second = hertz		
cal	calorie	H	henry
*cc	cubic centimetre	h	hour
cd	candela	ha	hectare
*cfm	cubic feet per minute	hp	horsepower
*cfs	cubic feet per second	*hr	hour
CHU	Celsius heat unit	Hz	hertz
Ci	curie		
cm	centimetre	in	inch
cu.	cubic (e.g. cu. ft = ft^3)	IT cal	international steam table calorie
cv	cheval-vapeur		
cwt	hundredweight		
d	day	J	joule
dB, *db	decibel		
deg	degree (of temperature difference)	K	kelvin
		°K	degree Kelvin
dyn	dyne	kg	kilogram

134

LIST OF UNIT SYMBOLS

kgf	kilogram-force	s	second
km	kilometre	sb	stilb
kn	knot	*sec	second
kp	kilopond	*sq	square (e.g. sq. ft = ft^2)
		sr	steradian
L	lambert	St	stokes
l	litre		
lb	pound	T	tesla
lbf	pound-force	t	tonne
lbm	pound mass	tonf	ton-force
lm	lumen	tonm	ton mass
lx	lux		
ly	light-year	V	volt
m	metre	W	watt
min	minute	Wb	weber
ml	millilitre		
mm	millimetre	yd	yard
mol	mole		
*mph	mile/hour	μ	micron
Mx	maxwell	Ω	ohm
		°	degree (angle or temperature)
N	newton		
nt	nit	′	minute of angle = $\frac{1}{60}$ degree
Oe	oersted		
oz	ounce	″	second of angle = $\frac{1}{60}$ minute
ozf	ounce-force		
ozm	ounce mass		

(′ and ″ were also often used for foot and inch, respectively)

P	phon; poise		
Pa	pascal	Symbols for Prefixes	
pc	parsec	a	atto	10^{-18}
pdl	poundal	c	centi	10^{-2}
Pl	poiseuille	d	deci	10^{-1}
*psf	pounds-force/square foot	da	deca	10
*psi	pounds-force/square inch	f	femto	10^{-15}
		G	giga	10^9
R, *r	roentgen	h	hecto	10^2
°R	degree Rankine	k	kilo	10^3
rad	radian	M	mega	10^6
rev	revolution	m	milli	10^{-3}
*rpm	revolutions/minute	n	nano	10^{-9}
*rps	revolutions/second	p	pico	10^{-12}
		T	tera	10^{12}
S	siemens	μ	micro	10^{-6}
*S	stokes			

Index

Note: In general, this index does not include dimensionless parameters (which are alphabetically listed on pages 96–121) or abbreviations (which are listed on pages 134–5).

abampere, 32, 46
abcoulomb, 32, 46
abfarad, 46
abhenry, 46
abohm, 32, 46
abvolt, 32, 46
'acceleration due to gravity', 12
acre, 43
air, properties of, 128
amount of substance, 20
ampere, 33, 34
ampere-turn, 127
a.m.u., 19
angle, dimensional formula of, 72
ångström, 123
apostilb, 37, 47
are, 123
astronomical unit, 123
atmosphere, 45, 130
atomic mass unit, 19
Avogadro constant, 20, 128

bar, 41
barn, 123
barye, 130
base units, 13
biot, 32
bit, 130
Bohr magneton, 128
Bohr radius, 128
Boltzmann constant, 128
British thermal unit, 30, 45
Buckingham's method, 63

calorie, 29, 45
candela, 36
candle-power, 36
carat, 130
Carnot cycle, 24
Celsius heat unit, 30, 45
Celsius scale of temperature, 28
centigrade heat unit, 30

centigrade scale of temperature, 28
c.g.s. electromagnetic units, 31
c.g.s. electrostatic units, 31
c.g.s. system of units, 17
charge on electron, 128
chemical similarity, 91
cheval-vapeur, 22
Child-Langmuir law, 84
coherent set of units, 18
convection, natural, 76
conversion factors, 42
correlation, spurious, 87
coulomb, 34
curie, 130
cusec, 130

dalton, 19
darcy, 130
day, sidereal, 130
  solar, 130
decibel, 130
defining equation, 50
degree (angle), 72, 124
dependent variable, 56
derived magnitudes, 49
derived units, 13
differential coefficients, nature of, 3
dimensional analysis, 54
dimensional analysis for directional quantities, 71
dimensional formulae, 50
dimensional homogeneity, 2, 52
dimensionless parameters, 93
dimensions, 51
dioptre, 130
direction, 5
distance, 4
dynamic similarity, 91
dyne, 17, 21

elastic similarity, 91

137

INDEX

electric quantities, dimensional formulae of, 79
electromagnetic similarity, 92
electron, charge, 128
  mass, 128
  radius, 128
electron-volt, 128
energy, units for, 22
erg, 22

f number of lens, 130
Fahrenheit scale of temperature, 28
farad, 35
faraday, 126
Faraday constant, 128
fathom, 131
fermi, 131
foot, 15, 43
foot-candle, 36, 47
foot-lambert, 37, 47
force, 6
frame of reference, 7
franklin, 31
friction in pipe, 59, 65
fundamental magnitudes, 48
fundamental units, 13

g, 11
gal, 131
galileo, 131
Galileo's Law of Inertia, 6
gallon, Imperial, 21, 44
  U.S., 21, 44
gamma, 46
gauss, 32, 46, 127
'Gaussian system of units', 33
General method (of dimensional analysis), 63
geometric similarity, 90
gilbert, 46, 127
grade (unit of angle), 131
gram(me), 16, 21
gram-force, 21
gram weight, 19
gravitational attraction, 6
gravitational constant, G, 128

heat, units for quantity of, 29
hectare, 43, 123, 131
henry, 35
hertz, 39
horsepower, 22, 45
hundredweight, 21

Imperial Standard Pound, 16

Imperial Standard Yard, 15
inch, 43
'incomplete' formulae, 54
independent variables, 56
indicial (Rayleigh's) method, 56
inertia, 6
inertial frame of reference, 8
integrals, nature of, 3
International electrical units, 34
International MKSA units, 34
International Practical Temperature (IPT) scale, 28
International Prototype Kilogram(me), 16
International Prototype Metre, 14
International Standard Pound Mass, 16
International Standard Yard, 15

joule, 22, 29

kelvin, 27
Kelvin absolute temperature scale, 27
kilogram(me), 16, 21
kilogram-force, 17, 45
kilogram-mole, 20
kilogram weight, 19
kilomole, 20
kilopond, 45, 131
kinematic similarity, 90
knot, 43, 131

lambert, 37, 47
light-year, 123
line of action, 5
line (of induction), 127
linear displacement, 4
litre, 21
long ton, 21
lumen, 36
luminous intensity, units for, 36
lux, 36

magnetic permeability of free space, 128
magnetic quantities, dimensional formulae for, 81
magneto-fluid dynamics, 85
magnitude of quantity, 1, 48
magnitudes, derived, 49
  fundamental, 48
  primary, 49
  secondary, 49
mass, 7
maxwell, 127
measurement, definition of, 48
metre, 14

INDEX

metre-candle, 36
'metric system', 14
mho, 126
micron, 131
mile, 43
mol, 20
mole, 20
momentum, 7

nautical mile, 43
neutron, mass of, 128
newton, 17, 21
Newton's First Law, 6
   Second Law, 7, 9
   Third Law, 8
nit, 127
numeric, 1

oersted, 32, 127
ohm, 34, 35

parsec, 123
pascal, 41
pendulum, simple, 56, 74
permeability, magnetic, of free space, 128
permittivity of free space, 128
phon, 131
phot, 127
'physical algebra', 2
physical similarity, 89
'Pi' theorem, 63
Planck constant, 128
planetary motion, 58
poise, 45, 125
poiseuille, 125
poundal, 17, 21, 45
pound-force, 17, 21, 45
pound mass, 16, 21, 44
pound-mole, 20
pound weight, 17, 21
power, units for, 22
prefixes, 15, 39
pressure, units for, 41
primary magnitudes, 49
primary units, 13
pro-basic set, 64
proton, mass of, 128
pseudovectors, 5
'puff', 126

quantity calculus, 2
quantity, physical, 1, 48

rad (unit of radiation dose), 131

radian, 39, 44, 72
Rankine scale of temperature, 28
rationalized equations, 35
rayl, 131
Rayleigh's method, 56
reciprocal poise = rhe, 131
recurring set, 64
reyn, 131
rhe, 131
roentgen, röntgen, 131
Rydberg constant, 128

scalar quantities, 4
scale effect, 92
scale factor, 90
scatter, 87, 88
screw propeller, 66
second, mean solar, 16
secondary magnitude, 49
secondary units, 13
sense (of direction), 4, 5
short ton, 21
S.I. units, 38
siemens, 126
similarity, 89
similarity of electric circuits, 91
slug, 18, 21, 44
solid angle, dimensional formula of, 75
speed, 5
sphere falling through fluid, 69
standard temperature and pressure, 128
statampere, 31, 46
statcoulomb, 46
statfarad, 31, 46
stathenry, 46
statohm, 46
statvolt, 46
Stefan–Boltzmann constant, 128
steradian, 39
stilb, 127
stokes, 46, 125
stress, units for, 41
Système International d'Unités, 38

temperature, 22
temperature scale, 23
temperature, thermodynamic definition of magnitude, 24
terminal velocity, 69
tesla, 35
therm, 30
thermal quantities, dimensional formulae for, 75
thermal similarity, 91
thermionic valve oscillator, 83

thermometer, constant-volume gas, 25
ton-force, 21, 45
ton mass, 21, 44
tonne, 21
torque, units for, 22
torr, 131

unified atomic mass unit, 19
unit, 1

vector algebra, 5
vector quantities, 4

velocity, 5
volt, 33, 34

water, properties of, 128
watt, 22, 35
weber, 35
weight, 10
weight per unit mass, 12
wire, twisting of, 72
work, units for, 22

yard, 15, 43